water chemistry for advanced aquarists

by guido huckstedt

Distributed in the U.S.A. by T.F.H. Publications, Inc., 211 West Sylvania Avenue, P.O. Box 27, Neptune City, N.J. 07753; in England by T.F.H. (Gt. Britain) Ltd., 13 Nutley Lane, Reigate, Surrey; in Canada by Clarke, Irwin & Company, Clarwin House, 791 St. Clair Avenue West, Toronto 10, Ontario; in Southeast Asia by Y. W. Ong, 9 Lorong 36 Geylang, Singapore 14; in Australia and the south Pacific by Pet Imports Pty. Ltd., P.O. Box 149, Brookvale 2100, N.S.W., Australia.
Published by T.F.H. Publications, Inc. Ltd., The British Crown Colony of Hong Kong.

Contents

ANALYTICAL SECTION

APPENDIX

FOREWORD

A knowledge of chemistry is vital to all aquarium owners, as even day-to-day maintenance requires an understanding of such important basics as pH and hardness. Yet there are some aquarists who feel that maintaining a certain balance of water chemistry will enable them to perform miracles.

These people have failed to grasp a basic fact that is more important than any chemical facts or figures. A knowledge of techniques may be easily obtained from books and other aquarists, but practical experience can be gained only through time and exposure to the problems involved in aquarium maintenance. It is the combination of book knowledge and experience that will make you a successful aquarist.

The Metric System

This book uses the metric system, which is a more international method of measurement than the English system of inches and feet. One meter (m) is equal to 1000 millimeters (mm). A centimeter (cm) contains 10 millimeters and is 1/100 of a meter. One inch is equal to about 2.5 cm or 25 mm. Roughly, 4 inches are about equal to 100 mm; 6 inches = 150 mm; 1 foot = 305 mm; 3 feet = 914 mm. The number of millimeters divided by 10 gives the number of centimeters.

As to liquid measures, the liter (l) is commonly used. One liter is slightly less than one U.S. quart. A liter contains 1000 milliliters (ml). The unit cubic centimeter (cc) is approximately the same as one ml.

To convert degrees Centigrade to degrees Fahrenheit, multiply degrees Centigrade by 1.8, and then add 32. (°C × 1.8) + 32 = °F. Good base temperatures to remember. 0°C = 32°F; 10°C = 50°F; 20°C = 68°F; 30°C = 86°F; 100°C = 212°F.

INTRODUCTION

This book does not contain the numerous tables all too frequently found in aquarium literature. Such tables are, admittedly, a convenient orientation for the reader, but in practice they are plainly misleading. For instance, the toxic effects of a substance do not depend on only one or two factors which can be described in a table. Example: In soft water, aluminum is poisonous; in hard water, it is non-toxic. Whether water is "soft" or "hard" does not tell anything about aluminum's possible toxicity; carbonate content is the determining factor. At a total hardness of $0°DH$ (read as "zero German degrees of hardness") enough sodium carbonate can prevent toxic effects—even though sodium carbonate is not included in hardness.

Tables paralyze the user's common sense. They successfully prevent observation and constructive thinking and can lead to failure if the reader regards them as something inflexible. Another example: There are tables on the pH tolerance of many different fishes; according to these tables, the ability to tolerate a certain pH depends on genus, species, size, and the general condition of the fish. But pH only indicates the degree of acidity or alkalinity of water, not the reasons why one fish thrives and another dies in the same aquarium. The more hydrogen (H^+) ions present, the more acid the water; the more hydroxyl (OH^-) ions, the more alkaline or basic the reaction. At a pH of 7, H^+ and OH^- ions occur in equal amounts. This is interesting and of practical importance, too, but the fish does not "taste" it. The fish is not "interested" in H^+ ions, but rather in what they "mean" in terms of anions (non-metallic ions). If the Cl^- (chlorine) anion is present, HCl (hydrochloric acid) and its H^+ cations (cations are metal ions) cause the acid reactions of the water. If the H^+ cation is combined with NO_2, the nitrite anion, the pH may be the same but the effect on the fish is totally different.

It is unfortunate that in the past aquarium chemistry has been *too* chemical—it has not paid enough attention to the importance of the biology of the aquarium. Although pH is measured with indicators, the value obtained is never more than an indicator itself. It is not absolute. Thus a pH of 4, caused by some mineral acid (hydrochloric acid, sulphuric acid, etc.), is dangerous; exactly the same pH, caused by organic (fertilizer) acids, not only does not constitute any acute danger, but may often even be an advantage.

The toxicity of heavy metals (with the exception of the nontoxic iron ion) varies considerably. Large quantities may be completely harmless, or traces of exactly the same metal salt can lead to a quick death. Here again quantitative classification would plainly be misleading, as the quantities given would only make sense in connection with a full analysis. This is no exaggeration! We have to bear in mind that the chemical method of examination reveals how many of the ions being sampled for are present in the test sample, but not necessarily how many are found in the original source, the aquarium water. If a 50 cc (cc = cubic centimeter = about ⅕ teaspoon) test sample contains 0.5 mg (milligram = 1/1000 gram) of ammonia, the water, to a chemist, contains 10 mg per liter. Mathematically this is correct, and in many instances physiologically as well. From a purely chemical point of view, this conversion is quite satisfactory, especially since this is the kind of result the chemist is interested in. The same is true of analysis for any other ions. But in aquarium chemistry we have to realize that the amount found in the sample, converted to quantity per liter, may not necessarily tell us the amounts present in the tank. The chemist must often change an ion from one form to another before he is able to measure it, so we, therefore, have to distinguish between the analytically established content and the *actual* quantity (= the quantity the fish is able to sense and react to).

This sounds complicated, but don't let it put you off! If you want to be on the safe side, you can eliminate a whole group of complications right from the start. Why distinguish between a total lead content and a temporarily bound lead content with the aid of complicated methods when there is not the slightest reason for using lead, including lead putties? This can really be said for all metals, even when, under the conditions in operation, they are not actually disrupting anything. Furthermore, it is important for the aquarist to rid himself of stubborn prejudices. The enthusiastic recommendations of lead-containing paints are slowly becoming tiresome and ridiculous. These paints are good where metals are to be protected from the influence of salt-free water—that is, they are excellent for garden fencing exposed to rain. As far as the aquarium is concerned, they are not only potentially toxic but do not last for any reasonable length of time. Synthetic latexes may be non-toxic, but they become brittle after a few years, and have been superseded by epoxy resins.

Finally, I wish to warn against unnecessary expense. This does not concern everyone since many aquarists feel happy only when they are able to use expensive equipment and can pursue their hobby amidst costly machinery. There is, of course, nothing we can do about that; snobbery is a way of life. The snobbish aquarist will undoubtedly laugh at the inexpensive ion-exchange device and determination of total salt content mentioned here, but that shouldn't worry us. But remember, it is important not to look at the simplified equipment as a cheap and dubious "substitute". Here lies a danger that I cannot emphasize enough.

From a certain stage onwards, no additional spending of money can buy the successful results of an expert approach.

All-glass aquaria are ideal for keeping marine animals. Potential toxicity resulting from possible corrosion of some areas of a metal frame is completely eliminated. Photo courtesy of O'Dell Mfg. Co.

1.
Oxygen

In the aquarium, a continuous consumption of oxygen (O_2) takes place. Not only do the fish use it up, but part of it is also lost through constant processes of oxidation. A process of oxidation occurs when a compound with no or little oxygen is changed into a compound with a high oxygen content. Thus the oxygen-free ammonia (a compound formed by one atom of nitrogen and three atoms of hydrogen, NH_3) is changed into the nitrate ion, a compound rich in oxygen (NO_3). By bacterial activity, the hydrogen was exchanged for oxygen: 1 atom of nitrogen plus 3 atoms oxygen then form one nitrate molecule, NO_3. The more processes of decomposition that take place in water, the greater are the oxygen requirements. The oxygen the fish require for breathing enters the aquarium through the aerator; consequently, the layman tends to let the air pump go full blast whenever there is a real or apparent oxygen deficiency. It is, however, very important to distinguish between real and false oxygen deficiency in order to decide what action should be taken. Furthermore, biological respiration is not the only cause of oxygen deficiency—non-biological factors, such as ion exchange or electron displacement, may also be responsible. In this case it is not primarily the gas household that has been upset, but there has been a disturbance of the electro-chemical equilibrium. Oxygen deficiency which has been caused electro-chemically can only be corrected effectively and with sufficient speed by the supply of atomic or nascent oxygen; aeration, with molecular oxygen entering the tank, is ineffective in this case. One has to choose from these two possibilities. The reader will regard the maintaining of the electro-chemical equilibrium as a very irritating introduction to a new and, what's more, complicated problem. What the aquarist would like to hear is the gospel truth: "A fish suffering from difficulty in breathing needs more oxygen—therefore, the air pump should be turned on." Unfortunately it is not as simple as that, but there is a simple practical solution. We have to bear in mind that oxygen deficiency may be confused with ammonia and carbon dioxide complications. In all probability most respiratory difficulties will be made worse by over aeration, and when there is no oxygen deficiency anyway, they definitely will because:

1. An already weakened fish uses up its last energy reserves fighting against strong water movement. Often the pump is turned on full and the fish are whirled about, the argument being that many fish have died of asphyxia but not one of a ride on the railroad.

2. Strong rotary currents whirl humus out of all corners. A roughly dispersed compost heap with a large surface forms, and is excellently suited for the adsorption of oxygen.

3. Owing to its slow reaction, molecular oxygen is useless where the oxygen deficiency has been caused electro-chemically.

USE OF HYDROGEN PEROXIDE

Generally speaking, a medication is not prescribed until the diagnosis has been made. There is, however, another method, called "ex juvantibus" by the medical experts, whereby the cause is deduced from the effective remedy. If the "patient" fails to respond, the cause will be a different one.

The aquarium owner has always been interested in the oxygen content of the water and what he misses most of all is a simple way of measuring it. In fact, only a rather complicated method is known to date, complicated above all because of the inconsistency of the reagents. I therefore propose an "ex juvantibus" method. It has the advantage of being completely harmless, and it acts instantly. To 20 l water, add 1.0 cc of a 15% hydrogen peroxide (H_2O_2) solution (= 1 drop per liter). This solves the whole problem of oxygen, temporarily, in a flash. The fish is allowed a recovery period of 10 minutes. If the symptoms have not markedly cleared up by then, the cause must, absolutely and without exception, be of a different nature. Where there is a slight improvement, an excess of ammonia is present or the

The amount of aeration in this marine tank is apparently sufficient. Oxygen depletion is easily recognized from distress shown by the fishes.

A very strong flow of air is decidedly unwarranted for the well-being of this fish. As discussed in the text, over-aeration seldom solves respiratory problems of fishes. Photo by G. Marcuse.

carbon dioxide content is too high. In water, hydrogen peroxide breaks down into oxygen and water. The speed of this process depends on the degree of purity of the water. With this method, all oxygen deficiencies of electro-chemical origin can also be corrected instantly and reliably because, when being formed, atomic oxygen is particularly reactive. The more reducing substances the water contains, the more quickly the nascent oxygen disappears.

2.
Carbon dioxide

Carbonic acid gas (H_2CO_3) dissolved in water is called carbon dioxide (CO_2). Most living organisms inspire oxygen and expire carbon dioxide. A very simple example can make this clear to us: about 20 cc of distilled water are put into a small dish and a few drops of universal pH indicator added. When checking the coloration against the color chart, a slight acid content caused by carbon dioxide in the form of carbonic acid will be observed. When a drop of tap water is now added, the pH will rise to about neutral. By blowing into the water through a straw, the water is quickly made acid again by the exhaled carbon dioxide. A similar process takes place in the aquarium. The fish expire carbon dioxide. If water movement were not provided, the fish would die of carbon dioxide poisoning and at the same time of oxygen deficiency as well. In aquarium practice it has become customary to interpret all respiratory disorders as oxygen deficiency. This is incorrect.

Aquatic plants, on the other hand, behave exactly the opposite of fish: during the light period they take up carbon dioxide and give off oxygen. This is why aquarists have always regarded plants as a source of oxygen. This, too, is incorrect. Each day is followed by a night and thus by more or less the reverse of the process described. As plants use up a lot of carbon dioxide during the day, a deficiency of carbon dioxide may occur, especially where the illumination is very powerful. This deficiency, as we shall see later on, can only be prevented under certain conditions, even if a lot of the water is changed.

An excess of carbon dioxide develops when too many fish are kept in one container and aeration is not sufficient to remove the excess. Carbon dioxide dissolves easily in water and is difficult to get rid of as the gas escapes from solution only when in

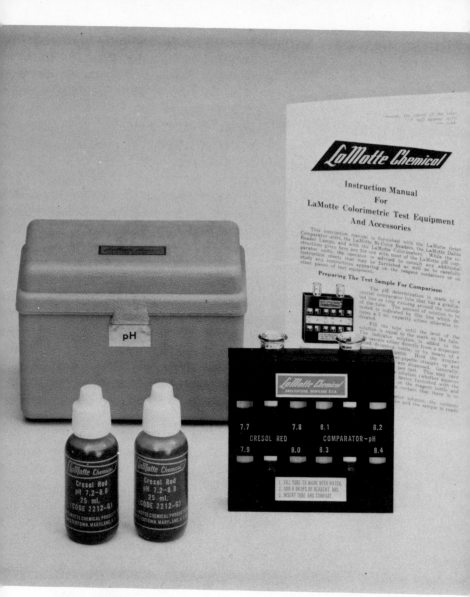

Measurement of the pH of aquarium water is made simple with the aid of kits available at pet shops. Photo courtesy of La Motte Chemical.

high concentrations. The main purpose of aeration is the removal of carbon dioxide. Its second, still important but not quite so extremely urgent, duty is the supply of oxygen. Oxygen is more easily added than carbon dioxide is blown out. In this context we would already like to mention that most peat varieties merely cause dirty water with a high carbon dioxide content. All peat should, therefore, be subjected to a brief examination (as described in the chapter "Filter Materials") before it is used for the aquarium.

Since carbon dioxide is closely related to pH, this is a good place for a review of pH, acids, and bases. Although water is usually written as a molecule, H_2O, some exists in the form of charged atoms called ions. Water constantly breaks down into positively charged hydrogen ions, written H^+, and negatively charged ions, OH^-. Negative ions have an excess of electrons, while positive ions lack their full share of electrons. If there is an abundance of H^+ ions in a solution, it reacts as an acid; if an excess of OH^- is present, the solution reacts as a base (also called alkali or hydroxide).

The pH scale ranges from 1 to 14, and is used to indicate the acidity or alkalinity of a solution. A solution of pH 7 is said to be neutral—that is, there are as many H^+ ions present as there are OH^- ions. Acidity is indicated by any pH less than 7; the lower the number, the greater the acidity. Any number over 7 indicates that the solution is alkaline. The higher the number over 7, the higher the alkalinity; pH 14 is the most alkaline condition measurable on the pH scale. A pH of 4 is considered very acid; one of 9 is very alkaline.

The salts of carbon dioxide are called bicarbonates (HCO_3) and carbonates (CO_3). They are differentiated by the alkaline reaction: bicarbonate solutions have a maximum pH of 8.4 - 8.6, the more strongly alkaline carbonates lie above this and, at very strong concentrations are termed alkalis. Hydroxides (i.e. bases or alkalis) can be present at very bright illumination when plants at first use up free carbon dioxide dissolved in water, then get CO_2 from bicarbonates, and finally even rob the carbonates of their carbon dioxide. The chapter "Standard water" describes how such changes can be successfully prevented.

17

If bicarbonate or carbonate solutions are made acid, an excess of free carbon dioxide develops. This process is known to everyone from effervescent powders such as Alka Seltzer. They consist of an acid (tartaric acid or citric acid) and a bicarbonate (usually called sodium bicarbonate or baking soda) which, after dissolving in water, release bubbles of carbon dioxide. If aquarium waters with a high bicarbonate content are made acid, the same phenomenon occurs and though not so clearly visible, numerous tiny bubbles can be seen. If aquarium water is, therefore, to be made acid so that the pH is lowered this has to be done in stages to allow the carbon dioxide produced to escape at intervals; otherwise carbon dioxide poisoning would inevitably result.

Together with ammonia, carbon dioxide is produced during processes of decomposition; it is, as already mentioned, an end-product of metabolism. For remedies see the chapter "Oxygen".

3.

Ammonia-ammonium

Even at low concentrations ammonia (NH_3) is poisonous to fish; the ammonium ion only becomes poisonous if present in very large amounts. Differentiation is easy with the aid of pH. The chemical differentiation is of less interest to the aquarist and shall only be mentioned. Ammonia is a water-soluble gas which combines with acids to form ammonium salts; ammonia is a molecule, ammonium is an ion. If the pH of a solution is changed by acids or alkalis, the ratio of ammonia and ammonium salts (if present) will also be changed: at the lower pH values, more non-toxic ammonium will be present; at higher pH values, more toxic ammonia will be liberated from the ammonium salts. We always measure ammonia in a strongly alkaline solution, and the toxicity of the amount present is established by measuring the pH and reading the table. The value arrived at may be helpful but is physiologically unreliable.

Tolerance to ammonia depends on:—

1. Type and size of the fish
2. General condition (starving and over-fed fish are more sensitive)
3. Carbon dioxide excess and oxygen deficiency (both increase the toxic effect)
4. The electro-chemical equilibrium, which will be discussed later
5. The colloid content (especially where sea-water is concerned).

Disregarding these factors or taking a favorable average value, the following quantitative interpretation can be made:—

Sensitive fish react at 0.2 mg with a noticeably increased rate of respiration, but more robust fish only react at 0.5 mg. At best, 2 mg can be tolerated for several hours without damage; at worst, 1 mg proves lethal within a short time.

The toxic effect of ammonia can very easily be confused with oxygen deficiency and an excess of carbon dioxide. This differentiation is of extreme importance and was discussed in the chapter "Oxygen".

Knowledge of the ammonia content is also important above all prior to each water change and to any chemical procedure. Almost all catastrophes blamed on water changes could be prevented by determination of the ammonia content. If the water change leads to an increased pH, and acid ammonium compounds were previously present, toxic ammonia will at once be liberated. The following table will be of help: —

pH-value	ammonia %	ammonium ion %
6	0	100
7	1	99
8	4	96
9	25	75
10	78	22

The table shows that at a pH of 7.0, only 1% toxic ammonia is present; 99% is harmless ammonium. At a pH of below 6.5, nitrification stops and the ammonia cannot be "broken down" by bacterial activity. There is, however, no actual destruction but merely a chemical reconstruction resulting in a removal of the toxicity. The very toxic ammonia is oxidized by the bacteria into the nitrite ion and then into nitrate, and thus rendered harmless. Above a certain threshold, very high nitrate content becomes dangerous again but we will not go into that here. Acid waters often possess a surprisingly high ammonium content, with 30 - 40 mg not exceptional. A water change of 50% may, therefore, result in acute ammonia poisoning if the old water contained sufficient ammonium ions to be changed into toxic ammonia by the raised pH. What to do in a situation like this is discussed in the chapters "Carbon dioxide" and "Oxygen".

Bacterial activity releases toxic ammonia, and some forms also attack the fins of fishes, causing fin rot, as seen in this blue gourami (*Trichogaster trichopterus*). Photo by R. Zukal.

Where do ammonia and ammonium come from?

Ammonia is the sign of organic pollution. It is a component of urine and feces, a product of albumin decomposition, and is formed from urea by bacterial activity. A more important source of ammonia is constant over-feeding, leaving food-remnants to decompose at the bottom of the tank.

4.
Treatment of abnormal respiration

When breathing periodically ceases and then proceeds with the opercula widely splayed, severe heavy metal poisoning is present. The signs of this, then, are irregular to interrupted respiration with very deep "breaths" after which breathing again stops. This must not be confused with occasional "taking a deep breath" during normal respiration. Take the fish out of the water *at once*. Only very healthy animals will survive even after being transferred to a different container. Optical differentiation of ammonia poisoning, carbon dioxide poisoning, and oxygen deficiency is only possible in very marked cases.

AMMONIA POISONING

Lowering the pH even one unit is sufficient. Add a few drops of 5% sulfuric acid, measure the pH, and add a few more drops. If the water has a high carbonate content, watch out for formation of carbon dioxide, and lower the pH gradually. When the pH has dropped by half a unit wait 15 minutes, and then drop it the second half. The same goes for sea-water, but in this case the pH can gradually be lowered to 7.5 and raised again a few hours later. All marine fish can tolerate a pH of 7.5 for a day without suffering any ill effects.

CARBON DIOXIDE POISONING

Fish recover almost immediately after the cause has been removed. Raise the pH one unit. At a pH of 8, an excess of carbon dioxide is practically impossible, and at pH 8.6 it is out of the question. Never raise pH when ammonia is present!

Improper water conditions can even occur in nature, resulting in a massive killing of fish populations, as shown here.

OXYGEN DEFICIENCY

For 20 l aquarium water, add 1.0 cc 15% hydrogen peroxide. This treatment also improves respiration slightly in ammonia and carbon dioxide poisoning. Should a marked improvement occur, administer for a short period only, for instance half an hour to two hours, then check for presence of atomic oxygen and, if required, add more hydrogen peroxide (see "Analytical Section").

The advantage of this oxygen therapy is that treatment is always carried out as if oxygen deficiency were present—afterwards one measures whether treatment was necessary. Action is taken first (under these exceptional circumstances!), thinking comes second. To check for the presence of molecular oxygen would be time-consuming and difficult, but checking for nascent oxygen is finished in no time at all.

Excessive dosage: Double the amount is harmless though unnecessary; only three times the stated dosage is of acute danger.

5.
Calcium/carbon dioxide balance

Hardness is a way of expressing the amount of carbonates (CO_3), bicarbonates (HCO_3), sulphates (SO_4), and other salts dissolved in water. We are mainly interested in the salts of calcium (Ca) and magnesium (Mg). Some salts, especially calcium and magnesium carbonates, are readily driven out of solution by heating or by a decreasing carbon dioxide content; these salts constitute *temporary* hardness. Other salts, such as sulphate, cannot be precipitated by heating, and comprise the *permanent* hardness. *Total* hardness is the sum of the temporary and permanent hardnesses.

Although hardness is measured in several different units, we will use mostly the German scale, or degrees hardness (DH). One DH is equal to 1 part of $CaCO_3$ dissolved in 100,000 parts of water. Another unit, ppm (parts per million) is also commonly used; 1 DH equals 17 ppm. Soft water usually indicates a hardness rating of less than 3 DH (50 ppm), moderate water about 6 DH (100 ppm), and hard water over 12 DH (200 ppm).

The calcium/carbon dioxide balance is present if the water contains only as many bicarbonates and carbonates as can be kept in solution by free carbon dioxide. Or, stated another way, calcium and carbon dioxide are in equilibrium if the carbon dioxide present in the water is sufficient to hold bicarbonates (and carbonates) in solution or to prevent them from being precipitated. This refers only to calcium and magnesium carbonates and bicarbonates responsible for water hardness. As opposed to sodium bicarbonate (baking soda) and sodium carbonate, they do not dissolve easily. Solubility increases with a rising carbon dioxide content and decreases when there is too little carbon dioxide and the temperature is higher. This is why aquarium heaters are always more-or-less covered with a layer of precipi-

Measurements obtained by this inexpensive water hardness kit are fairly adequate for the purposes of an aquarist interested in a range of values suited for keeping certain types of plants and animals. Photo courtesy Rila Products Corp.

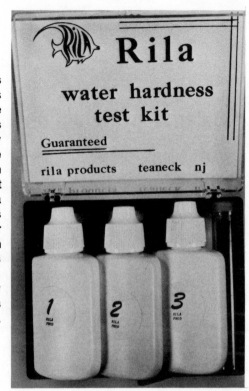

tated carbonates ("fur")—solubility decreases on the hot glass and leads to deposits.

If the balance of calcium and carbon dioxide is upset in the aquarium, the fur will be deposited where most carbon dioxide is being extracted. When free carbon dioxide is no longer available, fur will form on the aquatic plants themselves. This process of decalcification is also known as "biogenic decalcification". This expression is most convenient to the aquarist as he can now regard his plants, which are rotting away under a layer of fur deposits, as victims of a biological process. But if you love healthy plants and do not feel the urge to make animals suffer (pH jumps!), you would do better to use partial salt removal for decalcification purposes. Ion exchangers are far more efficient than aquatic plants, and can be regenerated as well. This point will be bought up again in the chapter "Standard water".

6.
Balance of reduction and oxidation

In the aquarium world, this is an as yet little known phase of equilibrium. I was the first to point out its importance for the growth of algae but at the time I merely wanted to point out, with the aid of sample figures, the mysterious appearances and disappearances of various kinds of algae "in one and the same water". The ominous factor X, which could be made responsible for this "self-willed" coming and going, had to be found. The botanists know very well what reasons to give for the importance of reduction potential, but these are mostly theories based on the field of cytology. During my intensive investigations I unfortunately failed to come across any practical material. It looked almost as though the practical influence of reduction potential on shape of growth, pigmentation, root formation, flowering, etc., had never been investigated systematically, presumably not even in conjunction with some other topic.

I can well imagine that a highly specialized worker could query and criticize some points of my argument. He has to bear with me. I am not an electro-chemical expert who has studied biology as a subsidiary subject and, therefore, had to compensate for the lack of expert knowledge by evaluating and reproducing numerous precise measurements. What compounds give off or take up electrons and under what conditions this exchange occurs has not been investigated in detail since, at least at present, this knowledge is not of practical importance in the aquarium field.

A simple process of oxidation has already been mentioned—the oxidation of ammonia into nitrate. Through oxidation, the reactive, oxygen-free ammonia is transformed into the more stable nitrate which is rich in oxygen. This process produces a barely measurable change of the electro-chemical equilibrium.

It is a different matter if part of the nitrate is reduced back into ammonia; in that case something is severely wrong with the reduction factors, as the reduction of nitrate is a phenomenon which accompanies severe reducing effects. An oxygen deficiency then occurs as another consequence of too low a reduction potential, and cannot be made up for by "full-blast aeration". The process of oxidation can thus cause oxygen deficiency, but so can the process of reduction. In both cases, however, oxygen may not be involved and, if so, any possible oxygen deficiency which still occurred would be a purely secondary side-effect. Water which is completely without oxygen may, therefore, have marked properties of oxidation, but can reduce just as well and as strongly when the oxygen content is high. In the former case, the hydrogen peroxide, forming atomic oxygen, would lower the oxidation threshold and at the same time correct the oxygen deficiency. In the second example, it would raise the oxidation threshold of the water and at the same time increase the normal oxygen content. Since the breakdown speed of peroxide depends on the reduction potential of the water, it acts as a virtual reduction buffer, because if we have water which is rich in oxygen, no excess of oxygen worth mentioning will develop; the atomic oxygen is slowly recombined into molecular oxygen. Let us summarize:—

Oxidation is either the adding of oxygen or the removal of hydrogen; reduction is either removal of oxygen or addition of hydrogen. If a substance is, therefore, oxidized through the removal of hydrogen, oxygen is not involved in the process of oxidation; this is also true where a compound is reduced by the addition of hydrogen. In both cases, oxygen can play a part in the oxidation-reduction effect, and in all cases, no matter how the components overlap, peroxide quickly causes a combination which is harmless for the aquarium.

Aquarium water differs from clean, natural water most prominently in the tendency of its electro-chemical equilibrium. In large or very fast running waters the tendency is always directed towards oxidation. In an aquarium, even in fairly large tanks with good water movement, the tendency is always directed towards reduction. In this respect an aquarium is remarkably different from a coral reef or a mountain brook, but there is little difference in the actual content of molecular oxygen, optical purity, or other striking effects.

The reduction tendency has a very great influence on what happens in the biological sphere, but the role it plays, even though so important, is at present still impossible to establish in detail. In this small volume we have to confine ourselves to the chemical and chemo-technical aspects. First of all, therefore, the main cause of tendency reversal should be known— bottom sand and filtration. Half the battle is already won if we allow the bottom layer to benefit from the circulation. To do this, we put a nylon-net-covered frame on the bottom and cover this with the bottom sand (not too fine) through which the water will flow from below upwards; usually a partial flow is already sufficient.

ELECTROLYTIC DISSOCIATION

The toxicity of many substances depends on the reduction potential. The ions present in water are electrically charged, i.e. they possess either one or several electrons more than protons, or vice versa. A molecule is considered electrically neutral because negative and positive charges are equal. Let us explain it this way: if those who are neutral and have no mind of their own are chased out of their warm nest and forced to live in another environment where they have to be on the move all the time, two parties will soon be formed and, according to the lessons history has taught us so far, one party will be "for" and the other "against". From time to time a few clowns will appear and found the party of the radical center—these are the molecules, and the whole set-up, in brief, is the principle of "electrolytic dissociation"—the break-down of molecules into ions. Continuous electron rearrangements take place in the water, and one almost gets the impression that one lot has nothing better to do than to steal electrons, while the others were fully occupied with trying to get them back.

Translated into scientific language, this means, "oxidation is the removal of electrons—reduction is the addition of electrons". When an oxidizing chemical is added to water, it is reduced (i.e., given some electrons) by the reducing chemical present, but at the same time the reducing chemical is oxidized (i.e., it loses some of its electrons to the chemical it reduces). This may sound like double-talk, but if you read it over carefully the concept is easy to grasp.

When electron displacement occurs, the chemical does not change, nor does it usually unite with another substance to form a compound, but its properties do change. Thus the toxicity of many substances depends on the reduction potential. One and the same substance may be non-toxic in the oxidized state and toxic in the reduced state. The reverse is possible, too, although it has so far never been observed in the aquarium. Simply by adding hydrogen peroxide as a source of atomic oxygen it is possible to remove the toxic effect of many compounds. Like all peroxides, hydrogen peroxide is an oxidizing chemical which forces its higher potential onto a toxic medium of reduction, gains electrons from it, and thereby renders it harmless. Should the toxic effect recur after a few hours, the toxic substance itself is a reducing system with a low potential. In a case like that one adds hydrogen peroxide and gets ready for a drastic water change. Where this possibility does not exist, the sad game can, as we know from experience, be continued for 6 weeks, after which the aquarium will still have to be vacated. But before that you should start using an ion-exchange device on the water.

7.
Theory of ion exchange

The term "exchange" is misleading for the layman, as it gives him the impression that only exchanges take place, perhaps of one salt solution for another. This actually happens in the so-called neutral exchange, but, firstly, the latter is undesirable in the aquarium and, secondly, this one special case does not characterize the principle of ion exchange.

If it were as simple as that, the process could be explained quite adequately on one page of the book and on the second page we could demonstrate its utter pointlessness as far as the aquarium is concerned.

The term "exchange adsorption" does more justice to the facts.

Let us say, for example, that we had to separate a mixture of rubble, gravel, and sand. Small amounts could be sorted manually; with larger amounts, we would do what they do on the building site. We put up a sifter and pass the mixture through it. The mesh width of the sifter (in the exchange granule the "matrix") depends on the amount of separation required. In all cases, only granules which are smaller than the mesh width can pass through the sifter. If all components of a mixture are to be divided separately, a whole set of sifters of varying mesh width are put up in a row.

This example can be transferred quite easily. With ion exchangers of varying pore-width, molecules and ions can be separated according to size, even if they are chemically similar. If it is the smallest portion of a chemical compound, a molecule, we talk of a molecular sifter. If we are dealing with electrically charged particles, ions, we use the term ionic sorting. Through calculation or measurements with an electron microscope, we know the width of the pores of the exchanger and can thus say in advance which particles are taking part in an exchange adsorption and which particles are excluded. To remove the quinine added to sea-water, we either use exchange adsorbents which adsorb the large quinine molecule only and let the ions pass through, or we use exchangers whose ability to exchange ions has been blocked so that only general adsorption is possible.

Although it is not quite accurate, we usually imagine any kind of salt to be an enormous collection of very many molecules. When salt is dissolved in water, the molecules disintegrate and ions form. Remember the example of the two parties. Within a party, too, we find weaker and stronger elements. Let us take one party to symbolize cations (these are ions with a positive electric charge); one party has a double electric charge, or even a treble positive charge. The other party is characterized by a negative charge (they are called anions); the weaker members only have a single negative charge, the stronger ones a multiple negative charge. In technical terminology, the expression "valence" is used. This valence is of importance for the aquarium, too (otherwise I would not trouble you with all these details), since the univalent ions have to be in some sort of sensible biological proportion to the bivalent ions. If this balance is radically upset, there will be disturbances.

We distinguish between cation exchangers and anion exchangers. This does not necessarily mean that cation exchangers only exchange cations and anion exchangers only anions. Technically, the two often overlap, but for aquarium usage we can forget about this and assume that the cation exchanger only exchanges cations, the anion exchanger only anions. Depending on the type of resin, both exchangers can adsorb molecules provided the diameter of the pores permits this; this happens with those resins which increase considerably in volume. Finally, on the exchange granules themselves there may be precipitations. By choosing the appropriate exchanger, any water can be changed as desired. The suitable combination of two types will give us distilled water.

8.
Ion exchange methods

NEUTRAL EXCHANGE

In this process the hardness factors, calcium and magnesium, are exchanged for sodium. It is an excellent procedure for laundries, hospitals, dye-houses, hairdressers—in short, for all institutions and any kind of business where large amounts of soap are used. The aquarist will not adopt this method unless he wants a water which is already bad anyway to become even worse. Fortunately this procedure is gradually becoming unpopular in the aquarium world. For the sake of completeness I would like to give a short description of the procedure.

The exchanger is prepared with sodium cations through being subjected to a strong sodium chloride (table salt) solution. If it previously contained calcium cations, the latter leave the exchanger and are replaced by sodium cations. After rinsing, the exchanger is ready for use: it removes the calcium cations from hard tap-water and replaces them by sodium cations. To be quite accurate, all the cations of a water are replaced by sodium cations. For comparison, let us look at a water we could find anywhere. Inflow (natural water) subjected to neutral exchange results in the following outflow:—

sodium chloride	sodium chloride
magnesium sulphate	sodium sulphate
calcium sulphate	sodium sulphate
calcium carbonate	sodium carbonate
magnesium carbonate	sodium carbonate

This means that the total salt content remains unchanged but also, above all, the carbonate content or the alkalinity. If the water was unfit to be used because of a high total salt content, it will, of course, remain so even after the procedure of neutral exchange. If the water was previously unfit for use owing to its high alkalinity, i.e., if it had a high pH, nothing will have been gained by this exchange. Sodium carbonates are just as alkaline as calcium and magnesium carbonates.

Another factor needs to be considered. Calcium and magnesium carbonates are very difficult to dissolve. As already described in the chapter on carbon dioxide, the solubility depends on carbon dioxide content and temperature. Sodium bicarbonate and sodium carbonate are not influenced by such factors. If aquarium water is, therefore, subjected to neutral exchange and evaporated water is replaced by exchanged water as well, there is nothing to hinder rapid and drastic rise in pH. If the carbonate content was already high from the beginning, the exchange may not push it up much more but it will increase further when the tank with its plant-growth is exposed to the sun-light, because, unlike calcium carbonates, sodium carbonates are, as we know, not precipitated.

In the aquarium, neutral exchange should, therefore, only be carried out if it is superfluous anyway, for instance with distilled water, as a pastime.

PARTIAL SALT REMOVAL

This procedure differs from neutral exchange in that it diminishes the total salt content. If the total salt content consists almost exclusively, or to a very great extent, of bicarbonates and carbonates, partial salt removal results, more or less immediately, in practically distilled water. For those hobbyists who look with disdain at simple devices such as ion exchange devices, partial salt removal can be more-or-less accomplished by a more difficult method. Heat a bucket of water for a long time, or boil it and let it cool. With more heat, carbonates and bicarbonates sink to the bottom. With this method it is just possible to more-or-less successfully carry out partial salt removal in a bucket of water. 200 1 water can be prepared quite comfortably with an exchanger bucket—not to mention that the chemicals required amount to less, in cost, than the gas consumption.

The only technical problem is that of regeneration, but this can be solved quite easily, A strongly acid cation exchanger is regenerated with 7 - 10% hydrochloric acid and afterwards rinsed with tap-water. The estimated requirements are 2 liters regenerative acid and 5 liters rinsing water per liter of resin.

Regenerative acid and rinsing water (the latter, of course, also containing a small amount of hydrochloric acid) flow into a plastic bucket and, before being disposed of, are neutralized with alkali. (If the strongly acid solution is poured into the bath, chromium plating and sealing will be damaged. Nor should the highly acid waste material of the regenerative process be poured down the lavatory as the sealing might be destroyed.) Regeneration with acid and neutralization of the acid remainder with an alkali will seem somewhat complicated to begin with, although it is really difficult to see why it should. Do not spill anything; wear rubber gloves if you want to protect your hands—especially the nervous and jumpy should remember this advice! When pouring the acid from the bottle into the bucket to be diluted with tap water, *always* hold the acid bottle inside the bucket and with the other hand cover the bucket with a sheet of newspaper; there is nothing interesting to see anyway. The process of dilution is so uninteresting that one might as well let the contents pour out of a covered bottle. Use the cover again when the alkali is added. With a simple sheet of newspaper, the whole process can thus be simplified and rendered harmless. After regeneration and rinsing the exchanger is in the so-called H-form, i.e. it is charged with H ions.

Citric acid can also be used to put the exchanger into the H-form if one regenerates with sodium chloride solution prior to regeneration with citric acid. This method takes little longer but is more expensive. Its main advantage is the avoidance of hydrochloric acid, which can be dangerous.

Thus we now have a strongly acid cation exchanger in the H-form. If we add some sodium chloride solution to this, it will transform all salts into the appropriate acids. If the water contains mostly bicarbonates and carbonates, the exchange process will produce mainly carbonic acid in amounts which correspond to the carbonate content. The greater part will escape very quickly and, if you prefer, the rest can be extracted.

USES

If the water contained carbonates only, one could produce completely pure distilled water by using the procedure of partial salt removal. But all natural waters contain other salts as well. The strongly acid cation exchanger transforms all salts into the appropriate acids. To make the process quite clear, it can be described as follows: —

Natural water (inflow)	through H-exchange results in (in the outflow)
calcium carbonate	through H-exchange results in carbonic acid
magnesium carbonate	through H-exchange results in carbonic acid
sodium carbonate	through H-exchange results in carbonic acid
calcium sulphate	through H-exchange results in sulphuric acid
sodium sulphate	through H-exchange results in sulphuric acid
sodium chloride	through H-exchange results in hydrochloric acid
sodium nitrate	through H-exchange results in nitric acid
calcium nitrate	through H-exchange results in nitric acid

Or, to explain it more briefly: —

Through H-exchange, all sulphates become sulphuric acid
Through H-exchange, all chlorides become hydrochloric acid
Through H-exchange, all carbonates become carbonic acid
Through H-exchange, all bicarbonates become carbonic acid
Through H-exchange, all nitrates become nitric acid

The exchange takes place strictly quantitatively—a fact we shall make use of later, for the determination of the total salt content. If other salts, apart from carbonates, are dissolved in the water, the pH of the outflow will be correspondingly low. If tap water contains a large amount of carbonates and only a little sulphates, as well as few chlorides (from sodium chloride), the outflow will not contain much sulphuric acid or hydrochloric acid either. A slight addition of natural water to this acid water would be sufficient for the neutralization of the free acid. The carbonates of the natural water then unite with the acids to form sulphates and chlorides. How much has to be added to the outflow to neutralize it depends entirely on the natural water (tap water) supplied. This mixture ratio determines whether the procedure of partial salt removal is of use. If an addition of about one tenth of natural water is sufficient to neutralize the acid outflow, conditions are ideal. This would equal a partial salt removal of roughly 90%.

If a quarter of a liter of natural water has to be added to three quarters of a liter of the acid outflow so that the neutral point is reached, the process of partial salt removal can still be regarded as fairly useful. The salt-removal factor is then not quite but very nearly 75%. We have to bear in mind that by adding natural water we not only get the carbonates required for neutralization but also neutral salts. Here, the exact salt-removal factor should already be established by the determination of the total salt content.

When neutralizing by adding natural water, the natural water is poured in small amounts and the pH is measured each time. Neutralization is completed as soon as there is a pH of 4.5 - 5. After brief aeration the pH would then rise to 7.

Let me stress this point: the mixture ratio is only measured once. Once we know that to one bucket of acid out-flow we need to add, say, one liter of natural water for neutralization, we keep this in mind and from then on ignore any slight variations of the tap water supplied. In some cities the tap water provided often varies considerably and, to be on the safe side, the pH should be checked. As this procedure can be carried out within seconds, it is not too much to ask. To make everything quite clear, let me quote, as an example, the process carried out with a typical water sample.

Inflow of natural water	*Outflow*	*Neutralized with natural water*
carbonate hardness 16	carbonic acid	carbonate hardness 0
sulphate hardness 2	sulphuric acid	sulphate hardness ±2
total hardness 18		total hardness ±2
pH, aerated, 8.4	pH 3.4	pH, aerated, 7.0
other salts: trace	other acids: trace	other salts: trace

From the aquaristic point of view, such water can be regarded as being completely free of salt, since only under exceptional circumstances can a fish tolerate a lower salt content.

MILDLY ACID EXCHANGERS

We must not omit to mention that partial salt removal can also be carried out with the so-called mildly acid cation exchangers. They specifically remove alkaline compounds only and multi-valent metal salts in particular. The advantage is that it would not be necessary to add natural water, but the disadvantages must not be forgotten. The mildly acid cation exchanger supplies completely purified water only at the very beginning. Resin is not able to cleave mineral salts, so it does not affect sulphates, chlorides, etc., and has, as is characteristic for these resins, a lower exchange speed. The strongly acid cation exchanger is simply hung under the tap, the tap is turned on and the water caught. If the cation exchanger is only mildly acid, the water has to flow very slowly so that a relatively similar result can be reached.

NATURAL WATER QUALITY

Generally, water along the eastern seaboard and southeastern coast of the U.S. are of good aquarium quality, with a total hardness of less than 3 - 5° D.H.; the same is also true of areas in the Pacific northwest. The central and southwestern parts of the country vary in water quality from locally excellent to absolutely terrible.

Below are average hardness data for four U.S. localities to help illustrate regional differences in water quality. 17 ppm (parts per million) equals 1° D.H.

Delaware River, Trenton, N.J.
Magnesium, calcium, and carbonates—under 200 ppm
Sulphate—under 100 ppm

Ohio River, Cincinnati, Ohio
Magnesium, calcium, and carbonates—500 ppm
Sulphate—300 ppm

Colorado River, Grand Canyon, Ariz.
Magnesium, calcium, and carbonates—1000 ppm
Sulphate—500 ppm

Columbia River, Rufus, Oregon
Magnesium, calcium, and carbonates—under 300 ppm
Sulphate—under 100 ppm

Chlorides parallel hardness by being more abundant in south-western and interior waters than in coastal drainages, at least those not near sea-water influence.

ANION EXCHANGE

This is of advantage when the method of partial salt removal is no longer sufficient. The additional use of an anion exchanger after a cation exchanger does not really imply any additional expenditure; just the original cost of equipment is slightly higher. The procedure will not take long to describe either since, in a sense, it is merely the "reverse" of cation exchange. First of all, let me explain the term "acid remainder": —

The acid remainder of hydrochloric acid is the negatively charged chloride anion (Cl) also known as chloride ion.

The acid remainder of sulphuric acid is the negatively charged sulphate anion or, in brief, the sulphate ion (SO_4).

The acid remainder of nitric acid is the nitrate ion (NO_3) of phosphoric acid the phosphate ion (PO_4), etc.

If, for instance, a highly basic anion exchanger carries chloride ions, it exchanges the anions of an added sodium solution for chloride ions. This would be exactly the reverse of the cation-neutral exchange. If a strongly alkaline bicarbonate-carbonate solution is added to a strongly basic anion exchanger in "chloride form", the exchanger removes these ions from the solution and replaces them with chloride ions. Thus we could also purify water containing carbonates by replacing the carbonate ions (acid remainder of carbonic acid), through anion exchange, by chloride ions. This would not change the salt content, but the pH would become neutral. If the strongly basic anion exchanger were present in the OH-form (regeneration with a mildly alkaline solution), it would transform all the salts of a solution into the appropriate alkalis. Thus sodium chloride would become caustic soda (NaOH), calcium sulphate would become caustic lime ($Ca(OH)_2$), sodium nitrate caustic soda, etc. That is, it does the exact opposite of the strongly acid cation exchanger which transforms all salts into the appropriate acids.

If we have a strongly basic anion exchanger in OH-form and add an acid mixture to it, it removes the acid remainders only

A variety of aquarium fishes have been successfully bred in out-door fish ponds of fish farms in Florida, even though the water in the pools may differ markedly from the properties of the fishes' home waters. Photo by Dr. Herbert R. Axelrod.

and replaces them with OH-ions. These negative OH-ions then unite with the remaining H-ions of the acid to form water.

If we, therefore, use a strongly acid cation exchanger in H-form and then a strongly basic anion exchanger in OH-form, we receive pure distilled water which, with a small buffer filter, can be adjusted to any pH between 5 and 8. If the two are mixed, even greater purity can be achieved. In this case, the strongly acid cation exchanger in H-form is mixed very thoroughly with the strongly basic anion exchanger in OH-form. The result is a strictly neutral water at least three times the quality of distilled water. After the mechanical mixture has been exhausted, both resins have to be separated by flotation and regenerated separately.

None of the possibilities described here are necessary for the aquarium. The examples given were simply examples used to give a clear description of the principle. The aquarium practice is different and simpler.

PRACTICAL REMOVAL OF TOTAL SALTS

In simplified practice, it is preferable to use either a mildly basic anion exchanger or a medium basic one. To do so has a number of advantages:—

1. A considerably greater capacity and a lower purchase price since a smaller quantity of resin is needed.
2. Maximum reliability. Strongly basic anion exchangers would be subject to disturbances and would have to be handled with great care.
3. Without requiring an additional buffer filter, mildly basic and medium basic anion exchangers could be used after the application of a strongly acid cation exchanger.
4. The resins are easier to regenerate.

With regard to water quality, the difference between strongly basic anion exchangers and mildly-medium anion exchangers is of no interest what-so-ever. The strongly basic anion exchanger also removes carbonic acid from the water and (if present) silicic acid. Due to some carbonic acid, the outflow of a mildly-medium basic anion exchanger causes a pH of 5.5 rising to nearly 7 after brief aeration.

The degree of purity arrived at with the aid of the mechanical mixture is much greater, but is of no practical value at all and is even extremely paradoxical. To keep the water pure in the aquarium, one would only be able to use paraffin-coated glass balls instead of sand for the bottom, since even the best quartz pebbles give off traces of silicic acid, which has previously been removed through filtration with the strongly basic anion exchanger. The glass plates, too, would have to be paraffin-coated, since even the best glass, etc.

Although the standard combination consisting of a strongly acid cation exchanger and a strongly basic anion exchanger is commonly used, its advantage can only seldom be utilized. The sensible aquarist will avoid playing such practical jokes on himself.

ION EXCHANGERS AS BUFFERS

Mild to medium basic anion exchangers only operate within the acid range; the added solution would, therefore, have to have a pH of below 5. Within the neutral to alkaline range, the exchange capacity is very slight. Some resins, on the other hand, possess an astonishing buffer capacity. The pH range of this buffering capacity varies from one resin to another. If we do not remain within the appropriate pH range, the buffer is still effective but its capacity has decreased and the exchanger would become exhausted sooner. Many resins thus have a "buffer interval".

The highest buffer capacity of the Lewatit MP 60, for instance, lies within the range of pH 8.1—pH 8.4. A resin of this type is regenerated with an alkaline solution (i.e. brought into the OH-form) and then rinsed with a highly diluted acid (as, for instance, produced by an acid cation exchanger) until a sea-water sample added meantime shows a pH between 8.1 and 8.4.* To keep sea-water at a constant pH of 8.3, the Lewatit MP 60 is adjusted to this pH and switched into sea-water circulation. The

---

* To reach the desired pH, any kind of acid water can be used. Should the first sea-water sample prove to have too low a pH, some sea-water treated with a carbonate-bicarbonate buffer is added.

An acid reaction of the water dissolves alkaline compounds present in most commonly used natural decorations and upsets the desired chemical balance. In addition, artificial decorations may also produce toxins of unknown composition in either a freshwater tank or in a marine tank like the one shown here. Photo by Alfred A. Schultz.

sea-water then remains at this pH of 8.3 with surprising persistence. If, for any reason, the pH of the sea-water drops (and it always tends to do this), the resin can neutralize the acid immediately with the alkaline OH-ions. To simplify the process, we will put it this way. If the sea-water remains unchanged, the exchanger has failed to act.

If the pH in a tank is to be kept at a constant level, the mildly acid cation exchanger Amberlite IRC-50 is excellently suited for this purpose. It is regenerated with diluted acid (or citric acid) and rinsed with tap water until the outflow has acquired the desired pH. This exchanger only takes up alkaline compounds and gives off acid H-ions instead. Apart from that, like all resins considerably increasing in size, it possesses the property of colloid adsorption.

With the aid of this resin, slightly acid aquarium waters can easily be kept free of ammonium ions. Although only a low percentage is removed per through-flow if the pH is low, this does not matter much if the resin is used all the time. This buffer resin does not cleave such neutral salts as sodium chloride, calcium sulphate, etc. Only when alkaline compounds get into the water does the exchanger begin to "act". As far as the control of the pH is concerned, its effect is exactly the reverse of the Lewatit MP 60.

In a solution, the Amberlite IRC-50 buffer thus acts against the alkaline side and Lewatit MP 60 against the acid side. Their greatest capacity lies within the direction mentioned here. If a water is to be kept constantly acid with Amberlite IRC-50, no carbonate compounds must be present in the water. The exchanger would, for instance, constantly remove calcium carbonates as they enter the solution. Although, with an intact exchanger, the pH would remain unchanged, the cleavage of carbonates would continuously produce carbonic acid and the resin would be exhausted relatively quickly. It would be strictly necessary to use quartz pebbles for the bottom of the tank and for filtration. This, of course, also applies if the pH is to be kept low without the aid of an exchanger. An acid reaction of the water dissolves soluble alkali compounds. The decorations (such as limestone, snails, snail shells, etc.) would have to be watched too.

CIRCULATING EXCHANGERS

In conclusion, a very interesting possibility shall be mentioned although I rather doubt it will be worthwhile for smaller tanks. Again and again an urgently required water change has to be postponed because the disturbance connected with it would be sure to disrupt some interesting, much longed for processes. Often, again, water which is still very good and low in salt has to be replaced because of "small" but physiologically quite effective quantities of nitrates which have spoilt it. This is bound to happen, for instance, when fascinating rearing problems are being watched in a very large tank.

With a filter consisting of a mechanical mixture, one can now carry out an indirect water-change. This will result in the purest distilled water, free of carbonic acid, and the whole combination can enter the circulation direct. With this method, the aquarium water is constantly diluted with the purest distilled water, enters warm and leaves practically equally warm. To preserve the salt content, some plaster and sodium chloride is added at intervals.

In our institute (Max-Planck Institut für Verhaltensphysiologie) we have now started to use a series of exchangers just for indirect water changes. The water change commonly used is now no longer required since this particular method is not only much cheaper, where a large aquarium is concerned, but also very much simpler. In this case we then use strongly basic anion exchangers, too, so that the out flow remains free of carbonic acid, but this is not essential. After the strongly acid cation exchanger in H-form, one could just as well use a medium basic anion exchanger in OH-form if the out flowing water can be aerated during the inflow so that the carbonic acid is removed. Looking at it from this aspect, we would recommend the following basic combination: first a strongly acid cation exchanger, then a mild or medium basic anion exchanger—these can be used by the hobbyist, too. The outflow should be fed into the tank via a small container with a powerful outlet. To be on the safe side, one should carry out twice as many anion exchanges as cation exchanges so that, if the anion exchanger is exhausted prematurely, the acid water of the cation exchanger does not flow in by mistake.

TECHNICAL CHOICE:

What resins one should purchase depends entirely on the purpose they are to be used for and on how much you are willing to spend. If partial salt removal is sufficient, merely a strongly acid cation exchanger is required. If you only work with fresh water and you need completely salt-free water, choose Permutit RS and Permutit E7P which are cheap. Where both fresh water and sea water have to be prepared, Lewatit S 100 or Permutit

RS should definitely be combined with Lewatit MP 60, because then the medium basic anion exchanger Lewatit MP 60 can be used, on its own, as an absorbent resin for the purification of sea water or, if necessary, for the removal of quinine from sea water.* If you want to afford the luxury of transparent plastic columns, you can use the indicator exchangers. They contain wash-proof color indicators; and by taking one look at the column you can check whether the exchanger is operating properly or exhausted. One only needs small amounts of the very expensive but excellent Amberlite IRC-50.

---

* With a 1 liter Lewatit MP 60, the quinine (added for therapeutic reasons) can be removed from 100 l sea water. Regeneration: Start by rinsing the colloids with diluted alkali, then separate quinine from the resin with about 1% hydrochloric acid + rinsing water.

9.
Ozone

Depending on how ozonization is understood and carried out, it is either a useful, highly skilled process or a usually harmless technical game.

Many already feel they are getting better at the mere mentioning of the name "ozone", especially if there was nothing wrong with them in the first place. There is the association, "An odor of cleanliness and freshness." Chemically and physiologically, this aspect is of little interest, although it is true that if ozone is present even in small quantities the air is pure. Where dust, bacteria and reducing substances of any kind are present, ozone is being used up. Thus the association mentioned above does contain a grain of truth, but only a grain.

If air which contains ozone is fed into the water, the highest oxidation threshold will be reached (provided the air contains sufficient ozone) long before there is free ozone in the water.

Many aquarists are under the misapprehension that ozonization means a high increase of the oxygen content. Not even a small ozone lamp is easy to tolerate indoors. The layman has to be forgiven for believing that a considerable odor equals a considerable quantity; one has to realize that this is not so. Even at a dilution of 1 : 500,000, ozone still has a distinctive odor. One will have to become resigned to the fact that a very small, even if strongly odorous, amount of ozone (O_3) can only provide a very small amount of oxygen.

After a life of 1,000 hours, a small ozone lamp had an output of not quite 4 mg ozone per hour; after 2,500 hours, it was merely a trace. Even when new, this lamp was hardly capable of making up for a small oxygen deficiency in a larger amount of water, however strongly it may smell.* Should there be an

---

* In the aquarium, complicated ozone lamps should not be used but rather ozonization devices with a constant output for permanent use.

Ozonizer (left) and protein skimmer. Photo courtesy Aquarium Stock Co.

oxygen deficiency of 1 mg per liter, the total deficiency in 200 l water would be 200 mg of molecular oxygen (O_2). But even if the lamp were able to replace this, it could still not produce so much oxygen that there would be an excess; there would merely be a normal oxygen content and an additional quantity of ozone. But in the aquarium this should not be allowed to happen.

With the aid of a fitted catalyzer one could, of course, reduce ozone to oxygen. But this method would not only be unproductive but also quite senseless, because minimal aeration supplies far greater amounts of molecular oxygen than one could possibly get from the traces of ozone by catalytic precipitation.

Let us summarize what we have said so far: the smallest amounts of very strongly smelling ozone only lead to the smallest amounts of oxygen; these are so minimal that they are of no consequence for the aquarium. Ozonization of the water is to be understood as being in no way connected with oxygen.

And now for the other side! Ozonization is a very effective and handy procedure (because the water is not being contaminated by chemicals) since it raises the oxidation threshold of the water to its limit, the highest potential. As we have already seen, a high reduction potential does not agree with an oxygen deficiency. To reach this effect, it is not necessary to carry the ozonization so far that ozone is actually present in the water. Through ozonization we reach values which are otherwise found only in the cleanest and purest waters of the world, the coral reef and the mountain brook. All reducing compounds are oxidized at the quickest speed; undesirable intermediate compounds, which often disturb the water for an uncomfortably long time and are the cause of critical complications, are soon made non-toxic; and the highly reductive reducing agents originating from many processes of decomposition and the very dangerous bacterial toxins are prevented from developing at all. In other words, dangerous complications are simply avoided. This, and this alone, is the real advantage of all ozonization. That the water becomes germ-free at the same time is not a typical effect of ozone—to make it germ-free, one can just as successfully use much weaker oxidation agents or even reduction agents.

DOSAGE

To determine the technical dosage is a problem. Here the small aquarium owner is at an advantage in that he can achieve maximum effects at a minimum of expenditure. If red algae do not disappear completely within a few days after treatment (and this goes for all brown algae and diatoms, too), it could be that the amount of ozone is insufficient or that the water has been totally messed up. One week after the brown and red algae have died, green algae should develop. If green algae fail to appear, the reduction potential is too low. We will then find the mixed zones with all three kinds of algae as described in chapter 16. A minor increase will then make way for green algae. On the other hand, however, a very slight decrease is enough to let loose the red algae!

The types of algae growing in a tank serve as an indicator of the water conditions. In the above photo the higher plants are obviously not thriving well under the conditions favorable to the lower plants.

CHEMICAL TEST

The practical worker will wish to know how long he can aerate with ozoniferous air before he is likely to cause any damage. This is of particular importance where smaller tanks are concerned. The chemical determination of the ozone consumption of the water is not difficult. A large amount of water, about 5 liters, is aerated with ozoniferous air. At intervals of 5-10 minutes a 10 cc sample is taken out and 7-10 drops of ortho-Tolidine added. If the samples become tinged with yellow after one minute, ozone is present. We then have to calculate how long it takes in the aquarium: if, for instance, 20 minutes of treatment were required for 5 liters of water, a 200 l tank will have to be aerated for 800 minutes. This is not quite accurate but gives us a good idea of what is needed. With regard to the aquarium, we also have to consider the contamination of bottom and filter, and a dead fish, unnoticed behind the corals, is enough to upset the whole calculation. Once a tank contains plenty of green algae, ozonization is no longer required; it would be completely unnecessary and, if done thoroughly, even dangerous!

Nitrites, in amounts usually present in aquaria, and hydrogen peroxide do not interfere with the o-Tolidine test. The iodine-starch test sometimes recommended for ozone is unsuitable, especially for sea-water.

10.
Total salt content

By total salt content we understand the sum of all salts dissolved in the water, irrespective what kinds of salt. Since, the more salts are dissolved, the better the water conducts electricity, the "electrolytic conductivity" can be taken as a standard by which to measure the total salt content. In aquarium circles, the bad habit of the total salt content being measured in micro-Siemens is just beginning to become fashionable. The significance of the total salt content is indisputable, since many fishes demand water with an extremely low salt content, especially if they are to be bred and not just "kept". It is, however, equally indisputable that, biologically, it is sheer quackery to add up all salts to form a total. To begin with it is not the total quantity which is of importance but—at least roughly—the kinds of salt. In theory (fortunately not in practice!) this would mean that the total salt content could only be interpreted physiologically in conjunction with a full analysis. In practice a full analysis is not required since, according to what is now known, a variation of 50% can be ignored. But a partial analysis cannot be avoided if the figure for the total salt content is to be more than mere showing-off.

In aquarium circles, tolerances of many hundred percent used to be quite the rule, and this mistake cannot be avoided simply by the purchase of a very expensive conductivity meter: the instrument is not in a position to supply the conscientiousness and that little bit of thinking the worker is unwilling to give. Physiologically, the determination of the total salt content is not reliable unless it is described roughly as follows:—

"Total salt content established by: Carbonate hardness . . . degrees; sulphate hardness . . . degrees; remaining salts, calculated as sodium chloride . . . mg/liter."

To put it like this means that the total salt content is divided into two groups: the group of bivalent ions (expressed in degrees of hardness) and the group of univalent ions as the remaining salt content. Whether the univalent ions, calculated as sodium chloride, actually consist of sodium chloride or univalent potassium compounds is quite immaterial, particularly since the nitrate ions can be determined separately.

To avoid ion antagonism of uni-or bivalent ions, univalent ions (sodium, potassium, chloride, nitrate) have to be in some sort of reasonable proportion to the bivalent ions (calcium, magnesium, sulphate). Protoplasm is able to expand or shrink. It is influenced by the chemical and electro-chemical properties of salts or their ions, such as their positive or negative charge, ionic radius, and their water cover. The expandable protoplasm is usually negatively charged and the positive cations, due to discharging, then have a shrinking effect. The reverse is true for the negatively charged anions. However, not only the positive or negative charging of the ion is of importance but also its strength or valence. Thus the bivalent calcium cation inhibits expansion more strongly than the univalent potassium cation. Thorough upsetting of the balance can, therefore, have striking biological effects which, as a rule, were not intended. The processes of expansion and shrinkage can be observed particularly well in an expandable exchanger if it is put into a column and solutions with various ions are added. The exchanger either shrinks or expands, often quite dramatically. In a lecture this process can be demonstrated with, perhaps, Amberlite IRC-50. If such an exchanger is in proportion with the external solution, the result will be expansion in layers which can be measured in centimeters and is determined not only by the pH and the concentration but also by the factors mentioned above. Fish eggs and plants react far more sensitively to expansion and shrinkage than do fish.

ION RATIO

The exact proportion of uni- and bivalent ions, with regard to their biological effect in the aquarium, matters very little, and a rule of thumb is sufficient. According to this, judging by my past experience, not many more univalent than bivalent ions should be present but a larger quantity of bivalent ions is permissible. If a higher total salt content is required, the simple addition of sodium chloride is not suitable. The correct method would be a combination of plaster and sodium chloride. In natural fresh waters the bivalent ions actually are in the majority. Where the equilibrium has been greatly upset, plant growth either ceases altogether or the plants become chlorotic, turn yellow and start to rot.

This happens with particular frequency in old waters with a relatively low salt content, where the nitrate content then greatly exceeds all other compounds. It is actually very easy to prove that the sole cause of this is ion antagonism. A stagnant tank can be livened up instantly if either the antagonistic salt is added to the water and the salt content is raised at the same time or if the proportion in one and the same aquarium is reversed through ion exchanges. Most hobbyists will tend to mix fertilizer into the sand; then they are surprised if, despite "best fertilization", plant growth still does not continue or becomes even more chlorotic.

11.
Standard water

It is less important to obstinately copy natural conditions than to allow natural factors to become effective. Whether "standard water" actually exists anywhere in the world and whether it is inhabited by fish and plants is completely immaterial, for the following reason alone: natural water is ruled by other factors, and this has already been shown in the chapter on biological equilibrium (oxidation-reduction). A complete reversal of the natural tendency was observed.

Aquarium water has to be made up in such a way that the calcium/carbon dioxide equilibrium exists under artificial conditions and that it is maintained whatever changes may take place. Fanatically nature-minded aquarists always become bitterly angry if an aquarium is subjected to completely natural occurrences such as, perhaps, a mass mortality. On the other hand, they can stand by and watch with a smile such gross disturbances as tremendous pH jumps and the rotting away of plants, as long as a joker humors them with a "biological excuse". It is "biogenic decalcification" I am speaking about. Although it occurs in natural waters, it does not happen in ideal aquarium waters.

The two characteristics of standard water are:—

1. A very slight carbonate hardness, no matter how high the sulphate hardness may be, and
2. A low total salt content.

The term "standard water" has been adapted from that of "standard earth". Standard earth has made the gardening practice a great deal easier and standard water has greatly simplified aquarium practice in our institute. For 4 years we have been using about 15-20 cubic meters of it per month, and over the course of the years I have dropped the habit of asking the permission of newly arriving fish, plants, and lower animals before putting them into it. As this patronizing behavior has not led to any mistakes, I now believe I am justified in generally recommending the application of standard water.

In possible exceptional circumstances, such as fish adapted to higher osmotic pressures, standard water allows us to add more salts as required. In this case the sulphate hardness is increased to 10-15 degrees and 100-200 mg sodium chloride are added per liter. Fish from brackish waters are exceptions anyway. A full analysis of standard water is really very simple.

In the following table, three figure values are given; the center row is the average value with the greatest frequency, the figures on its left and right are minimum and maximum values which have been tested very thoroughly.

Sulphate hardness	2	4	10
Carbonate hardness	0	0.5-0.8	up to a max. of 1
Chloride	2 mg	10 mg	100 mg

plus an additional trace mixture of iron citrate, phosphates and potassium.

BUFFER EFFECT

If we have completely salt-free water, the conversion table will tell us how much salt to add so that the desired content is reached. If part of the salt has been removed and the partial salt removal has not already proved satisfactory, only the difference is added. Slight carbonate hardness implies a very slight acid-binding capacity; to make the water acid, minimal quantities of peat are, therefore, sufficient. If water which is low in carbonates becomes old, the pH drops to 6, i.e. the buffer adjusts the water to this pH. The buffer effect is the result of an ammonium mixture. A carbonate hardness of 0.5 to a maximum of 1 degree agrees with a pH of about 7, and this pH remains constant and can only be lowered to the equilibrium of the ammonium buffer by severe organic contamination as in a rearing tank. The largest pH jump which could possibly occur is, therefore, one pH unit, and this only happens if the water is allowed to get really dirty. I admit that this cannot always be avoided where fish are actually being reared and it should not be regarded as a drastic mistake. In the rearing tank, too, ammonium is more or less non-toxic, whereas in a rearing tank with an alkaline reaction the very toxic ammonia constantly makes the fish gasp for air. If a shoal of fish-fry is divided up and one part is reared in a slightly acid, the other in a slightly alkaline pH range, one will be surprised to see how much more quickly the "acid fish" grow. This success can, of course, be achieved in slightly alkaline water, too, if the ammonia produced is constantly being removed. But whoever manages to do this in a rearing tank is clever and able indeed.

Let us remember that at a maximum carbonate hardness of 1 degree, the water is constantly in a stable calcium/carbonic acid equilibrium. This new proposition disagrees with the classical concept, but this is not a clashing of theories (both are correct) except that the classical theory leaves out several very important factors. The classical theory is a typical "desk-theory" in that it looks at the whole complex of "carbonate-bicarbonate buffering" in isolation. Let us now compare the barely buffered standard water with a so-called "well-buffered" water with a carbonate hardness of 3 degrees. In both cases we assume a content of 10 mg of free carbonic acid per liter, and in both cases we measure the pH changes which the removal of carbonic acid would bring about.

"Well-buffered water":—

free carbonic acid	10 mg	free carbonic acid	1 mg
carbonate hardness	3.0	carbonate hardness	3.0
pH	8.05	pH	8.05

"Standard water":—

free carbonic acid	10 mg	free carbonic acid	1 mg
carbonate hardness	0.7	carbonate hardness	0.7
pH	6.57	pH	7.56

This proves that the classical theory is correct: Through removal of carbonic acid the pH remains completely stable, whereas the pH of the standard water has changed by almost a whole unit.

And this is the argument against well-buffered water: if the removal of carbonic acid continues as through plant respiration, the pH jumps in two steps:

1. when the bicarbonate carbonic acid is removed,
2. when, after this has been used up, carbonate carbonic acid is affected as well.

This jump does not take place where standard water is concerned since aeration alone maintains the equilibrium. Standard water would only be less favorable if it were kept in a vacuum container. The above example is thus purely abstract: the figures are valid under the condition that no outside influences such as access to air upset the calculations. In practice, standard water dropped in carbonic acid content from 10 to 1 mg would, therefore, never make the pH jump established in the example; the actual jump is about 50% lower. The practical characteristics of standard water are minimal pH changes and good plant growth.

12.
Sea-water

First we must discuss a small problem, somewhat old but as important as ever: the pH determination of sea-water. Universal indicators correspond ideally within the measuring range of 5-7.5 because every half unit is represented by a different color. For the higher measuring ranges they are of no value and dangerous because of unreliability. Color shades like bluish-green, greenish-blue, blue tinged with yellow, and vague transitional forms cannot be read accurately, if they can be distinguished at all. Ten clear and, preferably, different colors simply cannot be arranged together. CZENSNY has added phenolphthalein to the Merck universal indicator and this at least prevents the reading of pH 10 as pH 8. Fairly adequate accuracy is also possible with Czensny's colorimeter. The ideal the practical worker still dreams of is a sea-water indicator with color differences as distinct in the pH range of 7.5-9 as is the universal indicator in the range of 5-7.5. To solve this problem for the sea-water aquarium, we tested numerous indicators and indicator mixtures. Finally, as sometimes happens, a quite simple mixture turned out to be more or less ideal. The procedure is outlined in the analytical section. One can make and calibrate the indicator oneself and control steps are not required.

RECIPE

Of sea-water recipes, the following is now commonly used. In 100 liters of water we dissolve the following:—

2765	g	sodium chloride, NaCl
692	g	magnesium sulphate crystals, $MgSO_4 + 7\ H_2O$
551	g	magnesium chloride crystals, $MgCl_2 + 6\ H_2O$
145	g	calcium chloride crystals, $CaCl_2 + 6H_2O$
65	g	potassium chloride, KCl

25	g	sodium bicarbonate, $NaHCO_3$
10	g	sodium bromide, NaBr
0.01	g	potassium iodide, KI
1.5	g	strontium chloride, $SrCl_2$

The nitrate addition sometimes mentioned is to be left out; furthermore, this recipe contains one gross error. There is not the slightest reason to replace calcium sulphate, which should really be present, with calcium chloride. Even the purest calcium chloride will give off free chlorine in solution. It is then necessary to let the water stand for a long period, to aerate it, or—in emergencies—to let it flow in through a charcoal filter. The water prepared according to the above recipe at first always has a slight chemical odor. Instead of calcium chloride, therefore, use 160-170 g of calcium sulphate, $CaSO_4 + 2 H_2O$, and save all the more-or-less unnecessary complications. The calcium sulphate is dissolved first; the water is made "milky" by it and becomes completely clear again after about 10 minutes. If precipitation occurs, large amounts of calcium are present; use only $\frac{1}{3}$ as much $CaSO_4$. Excellently suited is high spring water with greater carbonate hardness; in this case sodium bicarbonate usually doesn't have to be added. If, however, distilled water is used as the base, the amount of sodium bicarbonate stated is enough only in theory; the pH of 8.3 would only be reached after very long aeration, so add a few grams of sodium carbonate right at the start.

Often the recipe contains 5 g of primary phosphate as well. This quantity is not only unwisely high if added all at once but secondary phosphate is better anyway. About every other month a small amount on the point of a knife is added per 100 liters of water so that the phosphate is bound by the plankton. Iron is perhaps present only in the trace mixtures. It is not advisable to add iron in the form of inorganic compounds—it would just be precipitated straight away. Ferrous citrate is used, in the following way: a small amount—about the point of a knife—is dissolved in a little boiling water, stirred till dissolved, and added to 100 liters water.

A very good hint is to mix some fresh sea-water with a handful of garden earth, let it form a deposit, and pour the liquid into the tank. Repeatedly I observed that, during the first week, nitrification does not work in freshly prepared sea-water. Even if feeding is carried out with great care, there will be a build-up of ammonia. One could, of course, just as easily add old water, and in any case the advice only refers to a completely new tank, but it works excellently.

NITRATE

The great problem in sea-water aquaria is nitrate. All sorts of rumours are circulating about nitrate tolerance, and often one is "offered a choice" of very different values. The record was held by a very good tank, richly populated, and with "everything in best working order", with 1200 mg per liter. A good reference point are the data of the Aquarium Schönbrunn in Vienna where water samples were examined at regular intervals by professional chemists. The average there lies around 400 mg nitrate per liter, and the director of the aquarium assured me that the most sensitive new animals had been put into tanks with as much as 600 mg/l without coming to any harm. Quite certainly we can also believe the reports which lower the tolerance threshold to 250 mg/liter. In this case, the "truth" does not, as it is often supposed to, lie "in-between", but is found somewhere else. Above all, the quantitative theory of classification once again falls down. This leads to the conclusion that a certain nitrate content only becomes toxic under certain conditions. A simple case, observed quite by chance, was measured analytically and, again quite by chance, a catastrophe was prevented. Here, part of the nitrate was reduced to nitrite. The process of reduction always takes place in such a way that only a certain percentage of nitrite is produced from a certain amount of nitrate—the higher the nitrate content, the greater the percentage of nitrite produced. So far, a clear example of this has only been observed once, but doubtlessly there will be further opportunities.

What the sea-water aquarist desires most of all is an ion exchanger which specifically removes nitrates from sea-water. Theoretically, the matter itself is not really all that difficult, but one also has to make certain practical demands upon such a process. The reader should agree with the following: if nitrate

With the use of synthetic sea water it is no longer impossible for inland aquarists to keep marine invertebrates. Photo by R. Straughn.

is to be removed selectively (without any other changes occurring), one would have to save either time or money or, if possible, both, to make the procedure worthwhile. If the new method is not only more expensive but takes longer as well (complicated regeneration, etc.), and to such an extent that water change is much simpler and cheaper, then the procedure would be no more than chemo-technical playing-about.

But that is what the situation is like! Also, it would be much more sensible to remove the developing ammonia selectively, as was suggested for fresh water with the Amberlite IRC-50. This too can be done quite easily, but it is not worthwhile because the resins cannot be regenerated. Another, as yet not closely examined, possibility is the removal of nitrite, preventing the formation of nitrate. Theoretically, this succeeds with an exchanger blocked against sea-water and containing primary amines.

COLLOIDS

In the chapter "Ammonia" it was mentioned that much of the toxicity of sea-water depends on the colloid content. The colloid content is subjected to considerable variations, but the actual ammonia content (the amount the fish "is aware of") has only three possibilities.

1. The actual ammonia content is identical with the ammonia measured when no colloids are present.
2. The actual ammonia content is about 25, or at most 30%, lower than is indicated by the ammonia measured if a few colloids are present.
3. It is exactly 50% lower if a maximum of colloids is present.

Further lowering, by artificial adding of colloids (gelatins, casein, etc.), does not apply to the biologically useful range. The values mentioned above have been arrived at empirically. Where sea-water is concerned, do not, therefore, be surprised if one and the same fish under otherwise similar conditions, at an ammonia content of 1 mg/1, sometimes most frighteningly gasps for air and in other, more favorable, cases merely responds with an increased rate of respiration. The colloids found in sea-water aquaria are albumins which can only be detected according to instructions. They may suddenly, after the principle of "extraction", be deposited on the filter material, or perhaps on the bottom sand (slime coating) and any other kinds of objects. If larger quantities remain in solution, the water appears yellowish or bluish cloudy, depending on the color molecules adsorbed. Removal succeeds most elegantly with the resin adsorbent Permutit AS and the Lewatit MP 60 which are much more efficient than charcoal. But here, too, one will have to find out, by experimenting, if there are any exceptions. So far, the exchangers have proven greatly superior to charcoal, even though the exchanger becomes choked with mud; it can be completely regenerated later with diluted alkali. Afterwards, a desired buffer interval is created with acid. Resins such as the Permutit AS are also good, but, owing to the fine granulation, the filter resistance is usually too great.

General hints concerning the sea-aquarium can be found in WICKLER: "Das Meeresaquarium". The aquarium-chemical aspect naturally plays only a minor role in the book mentioned, as the book was written specifically for those interested in fishes. As I was informed by the author, the book contains all the popular recipes, even if, from the aquarium chemist's point of view, they cannot all be recommended without reservation. I hope you do not regard it as a contradiction, or lack of co-operation, if one author recommends recipes that another, from his point of view, has to criticize. As little as I would like to interfere with the work of the zoologist, the fish-specialist wants to lose himself in the changing of the recipes. "In an authorized way", such changes can only be suggested in aquarium chemistry, and they have to be based on sufficient evidence. So far then, there have been changes in the sea-water recipe: calcium chloride is replaced by calcium sulphate, the phosphates are used almost in the form of trace elements, and ferrous chloride, mentioned on p. 28 of "Meeresaquarium", is replaced by ferrous nitrate because, in sea-water, the latter remains in solution.

pH CONTROL

Now there is yet another amendment. The standard recipes for changing the pH of sea-water suggest raising the pH with sodium bicarbonate; sometimes sodium carbonate or alkali are recommended, too. For the lowering of the pH, phosphoric acid is commonly recommended.

First of all, I suggest a preliminary experiment. A small sample is mixed with sea-water pH indicator, and sodium bicarbonate is added. The pH drops in any case, no matter how much bicarbonate is added. From this we conclude that it is not very practical to raise the pH by previously lowering it. If carbonate or alkali are used, the pH can suddenly rise very quickly if it is not kept in check with the utmost care; then, of course, one has to lower it again with acid. Through a simple artifice, we can save ourselves from these antagonistic jokes: to six parts of sodium bicarbonate (U.S.P.) we add one part of sodium carbonate (U.S.P.) and all of this mixture (perhaps dissolved prior to use) is put in at once. In this way, the pH is buffered to 8.3 - 8.4. If too little was used, the procedure is repeated until a pH of 8.3 has been reached. One could, however, buffer beyond this by, for instance, adding a further 6 level teaspoons of sodium bicarbonate and 1 teaspoon sodium carbonate (to about 100 - 150 liter). With the above proportion, the buffer effect, or pH regulation, in sea-water is always produced at pH 8.3 - 8.4.

A knifefish (*Xenomystus nigri*) suffering from white spot disease or ich caused by *Ichthyopthirius*. Photo by Frickhinger.

Two-cell stage of *Ichthyopthirius*. This protozoan fish parasite is claimed not able to survive in acid water. Photo by Dr. H. Reichenbach-Klinke.

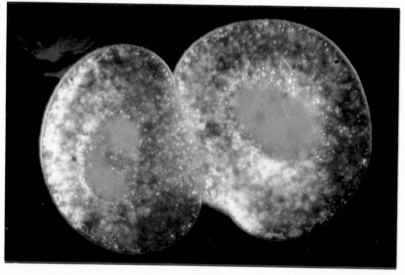

13.

Filtration

Looking at the numerous filter theories, one repeatedly gets the impression that all the aquarist is concerned about is the optical purity of the water. This is a secondary problem and, because it, too, is of importance, it has to be solved together with the main problem. Obvious dirt, leading to superficial diseases of the skin in fishes, is harmful to the plants, too, but it is nevertheless of secondary importance. Visible impurities are by no means the dangerous ones. The basic purpose of all filtering is the maintenance of the oxygen/carbon dioxide equilibrium together with simultaneous mechanical purification.

The numerous technical methods need not be described here, since all of them can be changed any old way provided the basic principle is adhered to. But this principle is already prevented from functioning when secondary problems become the main concern.

If one evaluates a filter according to its output per hour (how much water per unit of time it pumps through the filter material), he has already become a victim of self-deception. Only if the filter is extraordinarily efficient can it rectify the damage done by its inevitable side effects. This will be discussed later, but first let us take a closer look at the usual filters with an output of 100 - 200 1 per hour.

BOTTOM FILTERS

During the last 10 years one filtering method in particular has been carried out according to more and more pedantic and detailed instructions, and has been ever more highly praised in the absence of any reasonable foundation. In this method a group of internal filters sucks up the visible dirt through the bottom sediment towards the roots of the plants. There the dirt

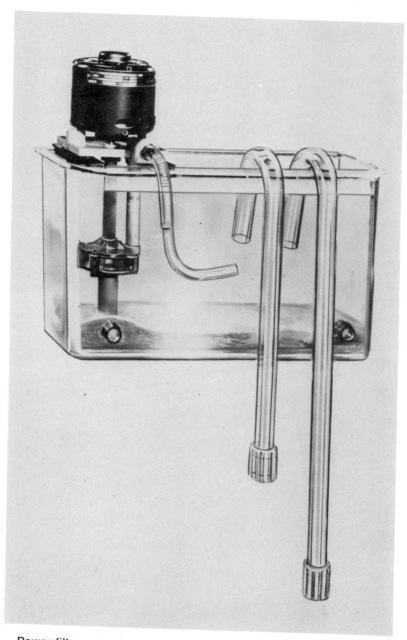

Power filters can pump more water than other types and are almost indispensable in large aquaria; they are especially useful in marine aquaria. Photo courtesy Eugene Danner Mfg. Co.

Labyrinth fishes like the chocolate gourami (*Sphaerichthys osphromenoides*) above and the croaking gourami (*Trichopsis vittatus*) below usually stay close to the surface of the water and are able to use atmospheric oxygen. Fishes with labyrinths exist with little difficulty in poorly oxygenated water. Photos by H. J. Richter.

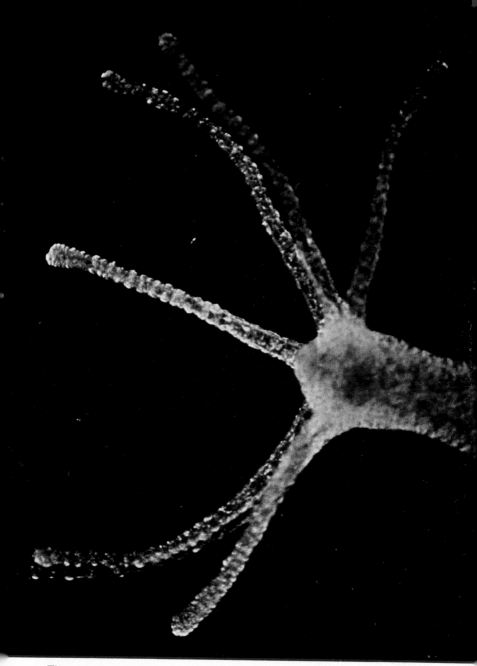

The presence of the common *Hydra* in the aquarium is not desirable. Some fishes may feed on hydras, but hydras can also paralyze fish fry with their nematocysts. Hydras can transmit some skin parasites of fish like the ciliate *Trichodina*. Photo by Dr. Reichenbach-Klinke.

is supposed to decompose quickly and—to use a laboured expression—"be broken down". This underground compost heap is advertised as a "depot of nutrients" for aquatic plants. Looked at from the view-point of plant physiology, this is sheer quackery. Aquatic plants are by no means "interested" in the depot; like all plants, they take up nutrients in the form of ions, not of particles. We do know of a secretion of plant roots, but this merely encourages the decomposition of carbon dioxide-soluble compounds. Organic fertilization has its advantages in gardening and agriculture as far as the so-called "cold soils" are concerned; through the dark coloration the soil becomes warmer. The so-called soil maturity also plays a part, and so does the structure of the soil. Here organic fertilizer exerts a favorable influence. But that is about all there is to it. There is no sense in improving the ability of aquarium earth to hold water by using organic fertilizers, since the bottom sediment in the aquarium only dries up under exceptional circumstances. What *is* of importance for the bottom layer of the aquarium is good ventilation; this will be dealt with later on.

OXYGEN DEFICIENCY

The filter is supposed to trap the dirt present in the water—dirty water is continuously sucked through the filter or filter material and optically clean water is given off. After a short operating period, relatively clean water is sucked through the filter and part of the oxygen dissolved in the water is used up for the oxidation of filter dirt. No matter how crystal-clear the water may be when it leaves the filter, its gas-balance has been seriously upset. The freshly filtered water is not only lower in oxygen, but it also has a higher carbon dioxide content. An oxygen deficit of 50% may develop. This well-known fact has frightened some theorists so much that they have become fanatical opponents of filtration. Their solution, "Away with all the harmful filters!", lacks refinement and is rather amusing in its indescribable naivety. It has already been mentioned that oxygen is easy to add but carbon dioxide is difficult to remove. All we need to do, therefore, is to make up for the damage the filter has done and, some way, bring back to normal the oxygen/carbon dioxide equilibrium of the water leaving the filter. The principle is quite simple: the pure water with its equilibrium upset has to be fed back into the tank via the water surface in as broad a flow as possible. Oxygen is quickly taken up, and the carbon dioxide escapes easily if the water flow is broad and dispersed or if the flow moves along the water surface. Dispersal of the flow is the most elegant method. If the flow is fed directly into the water, some carbon dioxide may leave at the moving water surface, but part of it will continuously be washed back into the water, as carbon dioxide is heavier than air and shows no tendency to rise above the water surface. All filtration is, therefore, aided by making sure that some air always has access to the water surface. Tightly fitting covers are bad and a suitable compromise will have to be reached. Some cracks should be left open so that the circulation of air present in every room anyway increases aeration. Where open external filters are used this is, of course, unnecessary.

If the flow is fed back into the water directly, not only will carbon dioxide be washed in continuously but the aquarium water will also be "diluted" all the time with water containing little oxygen. Fortunately oxygen quickly diffuses into the water from the surface and makes up for some, if not all, of this foolishness. If many fish are present in the water, its quality is dubious in a biological and chemical respect as well, as there is a slight oxygen deficiency all the time. To let this happen is not just foolishness but extreme stupidity! The occasional "blunder" is quite permissible, but the blunder should not become embarrassing.

OUTFLOW AND POWER FILTERS

Absurdity reaches its peak with the so-called outflow filters. Basically, they consist of a disguised capsule with holes—a shell, diving-bell, or some other obscure-looking object—inside of which an outflow is hidden. The dirt which has been sucked up is deposited in a depression of the capsule and the water in the tank is, from the bottom upwards, mixed with water which has a low oxygen content and a high carbon dioxide content. Apart from extreme cases, the quality of a filter does not depend on the output per hour. What is important is whether, together with the removal of dirt particles, a low carbon dioxide content can

A combination of factors including the composition of the water is considered the possible cause of little known fish diseases like the enlargement of the yolk sac of newly-hatched fish (above, photo by Dr. W. Foersch) and the accumulation of gas in the eye-ball of fish (below, photo by Dr. H. Reichenbach-Klinke).

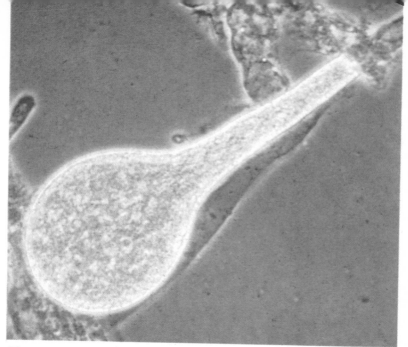

The spore-bearing capsule of the fungus *Saprolegnia* (above) under high magnification and a wounded *Scatophagus argus* with a secondary infection of *Saprolegnia*. This pernicious fungus develops mostly in neglected tanks, especially those in need of additional aeration. Photos by Frickhinger.

be maintained at the same time; the oxygen content will then be sufficient in any case. To check, 20 - 30 cc water are taken out of the tank in a clean glass and, with the air pump, aerated for a quarter of an hour. The pH is then measured and compared with the pH of the aquarium water. The aquarium water can be allowed to have a slightly lower pH; should the difference consist of a whole unit, e.g. pH of the tank 6.5 and pH of the aerated sample 7.5, filtration has to be adapted accordingly.

If the output of the filter is exceedingly high, as is for instance the case with power filters, disturbing factors are of little consequence because, quite apart from anything else, at an output of 400 1 per hour the period of contact between water and dirt is too short to permit any measurable oxygen consumption. On the whole, practically any problem of filtration can be solved with power filtration.

14.
Filter Material

What material is chosen is more or less a matter of taste as long as it fulfills its purpose. Basalt chips do not serve this purpose, however good they may be for a dark soil background. Glass wool is an old-fashioned filter material and it belongs neither in the household nor in the aquarium, as it is far too dangerous. Nylon wool has the advantage that it can easily be washed and used again. Gravel-filters are well-known for their suitability, with a mixture of fine and coarse gravel being of particular advantage. This mixture traps fine floating substances and the rough granules added to it help to prevent the material from collapsing and sticking together. The two filter media peat and activated charcoal have to be described in greater detail.

PEAT

The various kinds of peat are totally different substances. One quality may be completely useless, another highly effective, in a negative as well as in a positive sense. The opinion that all peats are acid is curiously common. Most peats are hardly acid at all, irrespective of their origin. As filter media, these barely acid peats are as useful as a frayed door-mat. For the aquarist it is important to examine the peat for its suitability to the aquarium.

1. Check the structure. Peat which still shows a natural coarse-fibred, matted quality is harmful for the aquarium, as it contains substances which inhibit plant growth. As opposed to this, crumbly peat contains growth substances and hormones, and usually fungicidal (fungus-destroying) substances as well, all of which promote plant growth, particularly root formation.

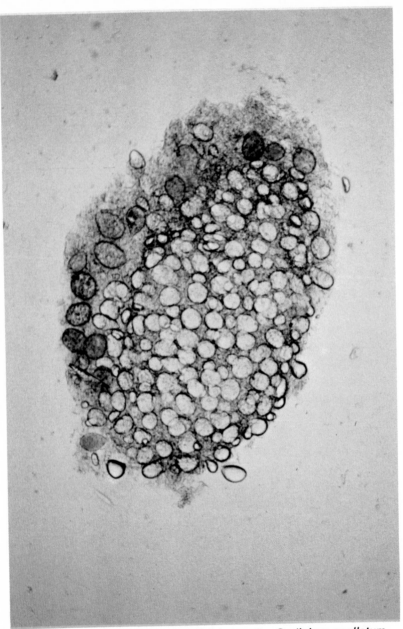

The introduction of the dinoflagellate parasite *Oodinium ocellatum* (left) into a marine tank is avoided by first isolating any new fish in quarantine. Photo by Dr. H. Reichenbach-Klinke.

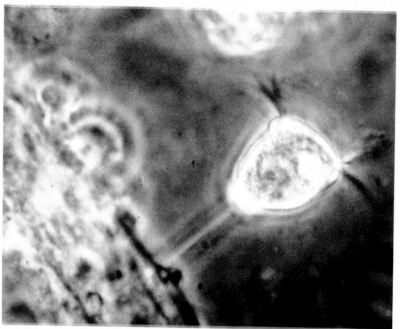

Large colonies of the free-living ciliate *Vorticella* can grow in stale tanks. They cover the plants and other exposed areas; there even are reports of their presence on the skin of fishes. Photo by Frickhinger,

2. Test for organic volatile acids. Soak one tablespoonful of peat in a tumbler of distilled water and let it stand over night. The next day, or the day after, enough water for the determination of the pH is taken out. If the sample is strongly colored, it may be safely diluted with distilled water (not tap water!). The pH must at least be below 5 and must not rise with aeration. If the pH rises due to aeration, the peat contains volatile organic acids and is unsuitable.

3. Test for humus acids. Peats with a high humus acid content usually stain water strongly and react markedly acid, with a pH of 3 - 4. In some acid peats the coloration is not very strong; color is not, therefore, a reliable guide. What matters is that acid peat-water maintains its pH even when the sample is subjected to aeration!

The humus acids found in acid peat are an excellent buffer

against the alkalinity; not even the most ingenious chemical buffer could take their place. Furthermore, these acid peats reduce oxidation.

ACTIVATED CHARCOAL

In the past activated charcoal has been accused of chemically changing the water, but now it has to be accused of unfortunately doing so no longer. Even very expensive varieties are presently hardly suitable for decoloration purposes! Another lot had an excellent decoloration effect but also a rather dangerous reducing effect, reducing most of the nitrate to ammonia. Other lots struck one with their enormous acid-binding ability, and the coal had to be treated with a lot of acid before use and the salt had to be laboriously washed out afterwards. These are only a few examples.

It is unfortunate that manufacturers no longer realize that for the aquarium, only media of constant quality can be used. The activated charcoal available on the market is no longer constant, and at present the use of charcoal means taking pot-luck. Perhaps one could get the manufacturers to agree to the production of a quality that remains constant, but a shorter, cheaper, and more elegant way is the employment of ion exchangers instead of charcoal.

15.
Biological equilibrium

The expression "equilibrium" does not require an explanation. The word "biological" preceding it presumably only means that different kinds of equilibrium can be understood as belonging to one complete process. It is somewhat difficult to say. The *term* "biological equilibrium" appears, in practice, to be of greater significance than the matter itself. In a manner of speaking, this term is used as a catch-all, and whatever seems suitable or happens to be in the way, the definable and the indefinable, is thrown into it. It is also excellently suited for covering all sorts of nonsensical statements. Some prominent theories of equilibrium are listed here, and it is up to the reader to decide whether to finally summarize them all under the adjective "biological". The biology flag which has been stuck on is not necessary, but it sounds interesting.

THE NITROGEN CYCLE

A state of equilibrium would occur if nitrogen products (ammonium, nitrate, urea) were to be taken up by the plants to the same degree as they are being supplied. This actually happens in large waters, but in aquaria it could only take place if the tank bore very little resemblance to an aquarium. In that case we would have a breeding tank for algae.

There is in fact such a thing as an algae tank, a very cleverly devised installation from which 70 g of algae albumin can be reaped daily per square meter surface. But these are no longer aquaria; in fact, a fish could survive in them only if harvesting were carried out every day—otherwise the dying algae, as soon as they were overgrown by new algae, would cause albumin decomposition. A device of this kind could not even be installed in the home. It requires the southern sky with large quantities

Corals like the stony corals growing above in a tide pool in the Fiji Islands require conditions that are very difficult to duplicate in the aquarium. Photo by Dr. Herbert R. Axelrod.

Away from their habitat, corals die and foul the water quickly; they usually are used only as decorations like the dead and bleached corals shown. Photo by Dr. D. Terver.

of light, a great deal of carbon dioxide, and, last but not least, the suitable types of algae.

When one is comfortably seated by the fireside, relaxed and meditating, pleasant thoughts will soon appear. One may arrive at crystal-clear conclusions, but there may also be a lot of day-dreaming. In fact, one usually has day-dreams. As long as aquaria have been in existence, the aquarium owner has had hallucinations of the nitrogen breakdown taking place in his miniature waters as well. Unless his eye happens to take delight in the interesting behavior of his fish, his look becomes misty and before his mental eye the algae tank begins to take shape, or the growing of aquatic plants, or perhaps—symbol of the eternal cycle—the snake biting its own tail. This dark metaphysical shudder, characteristic of a deep feeling of being one with the universe, is unfortunately deprived of its purpose by almost all aquarium owners and merely channelled into doing nothing: they not only regard a change of water as unnecessary but even consider it extremely dangerous! The "biological equilibrium" might be upset by it! In all cases, the existence of a plant or an algae is taken as evidence for the breakdown of nitrogen, but no one ever asks how much nitrogen is supplied and can actually be taken up again by the plants present.

ALGAE YIELD

The idea of fitting a strongly illuminated algae container to a seawater aquarium is a typical fireplace idea. Since it has already been tactfully condemned by STERBA in vol. 2 of "Aquarium science", there is no need to say any more about it. By chance I came across a rather brief, insignificant looking note by R. GEORGE in the DATZ which in its conciseness and clarity offers a very good definition. Of all publications on the nitrogen cycle in the aquarium, this brief note is by far the best : —

"Assuming that the biological equilibrium exists in a newly installed and planted tank and that 100 g of dry substance (food) are put into the tank per month, it is necessary, in order to maintain the biological equilibrium, to take out 100 g of chemically similar dry substance every month."

This convincing deduction is immediately understood by every botanist. Since this book is, however, designed for aquarium friends I shall supply the botanical data. Please do not regard it as an act of malice if I destroy a favorite, carefully nurtured illusion. It cannot be the duty of aquarium chemistry to darken metaphysical self-satisfaction.

Aquatic plants and algae consist of about 95% water; 3% of the fresh weight (without water attached to the algae!) consists of albumin and one sixth of the albumin content equals the nitrogen content (taken as N; as already mentioned, nitrate is a compound of nitrogen = N and oxygen = O). To make it easier to understand, let me quote a head of lettuce as an example. This may sound odd but is not out of context, as a head of lettuce has a water content of 95% (which equals the average water content of aquatic plants) and there are experimentally proven data available with regard to domestic plants. 1 kg of fresh lettuce contains 1.6 g nitrogen (= N) and 13.7 kg lettuce require about 22 g N to grow from seedling to being ready for sale, that is an amount of nitrate of 100 g NO_3. To maintain the nitrogen equilibrium in a tank we would have to harvest roughly 13 kg of green plant material per month if the tank has a capacity of 1000 l. For the smaller tanks commonly in use and holding about 200 l water, it would be one fifth as much, or 2.7 kg of green leaves.

It was taken that on an average at least 100 mg nitrate per liter are present per month. If one liter water contains 100 mg nitrate, a 10 l basin contains 1000 mg nitrate = 1 g nitrate, and a 200 l basin a total of 20 g!

One liter of wet filamentous green algae, thoroughly wrung by hand, produces half a liter of algae mass with a weight of 550 g, which would when dry only weigh 75 g. Half of this consists of non-nitrogen compounds such as carbon, organic acids, lipoids, etc. The other half of green leaf tissue is divided up into about 5% hydrogen, 30 - 40% oxygen, and 2 - 4% nitrogen; the remainder is ash with alkaline—and earth alkaline metals, phosphorus, sulphur, and trace elements. Taking a higher nitrogen content of 4%, one finds at best 3 g of nitrogen in half a liter of green algae. In an equal amount of blue-green algae hardly any nitrogen is found, since they consist of more than 99% water.

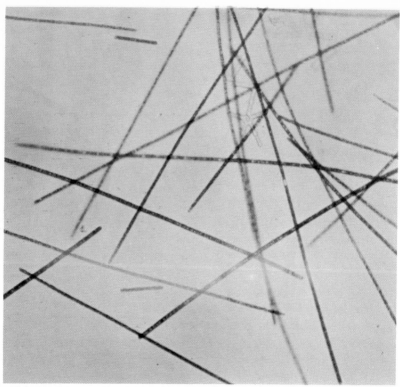

The blue green alga *Oscillatoria* above can move and thus cover everything in an aquarium with a blue-green film. Photo by Dr. H. Reichenbach-Klinke.

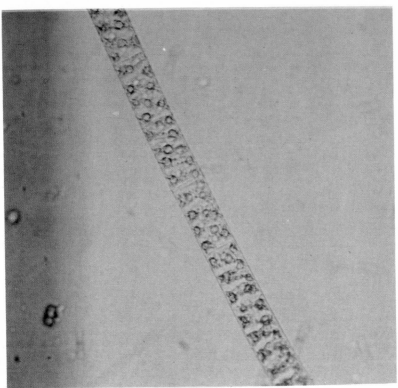

The green filamentous alga *Spirogyra* is one of the green algae that cause the green film in most tanks. Photo by Dr. Reichenbach-Klinke.

Finally, for the benefit of the beginner, we would like to point out that 1 liter of wet or half a liter of wrung algae in a 200 l basin already looks like a lot and doesn't grow in one day.

With a very good growth of algae the nitrate content of aquarium water can be lowered quite easily if the algae grow so quickly that they have to be harvested all the time—and this is particularly important—if they grow in amounts which happen only very rarely and only by chance in a room aquarium.

The situation is much more favorable, where freshwater is concerned, if an external filter, half the size of the tank, is planted with irises. Although they do not flower, they grow like weeds, look attractive, and really function. Every four weeks a few plants have to be pulled out to prevent the undesirable matting of the roots. The filter has to be placed near a lighted window, of course!

16.

Biological equilibrium and algae

LOW REDUCTION POTENTIAL

The botanist will quite certainly object that it is unbotanical to arrange algae according to their coloring matter. But here we are not concerned with botanical classification. We regard the differently colored algae as reduction indicators. They are in fact very accurate and reliable biological indicators, but unfortunately they do not respond quickly enough to be called ideal. Sometimes brown, red, and green algae occur together in sea-water; in that case optimal living conditions exist for none of the three. The undesirable brown algae and diatoms can be destroyed with hydrogen peroxide, an oxidation agent of medium strength. This raises the oxidation level to such an extent that soon a lively growth of red algae sets in and continues until all the red algae are destroyed with the next stronger oxidation agent, for instance ozone. Red algae contains carotenes and these are particularly susceptible to a raising of the reduction potential. Carotenes are anti-oxidants, i.e., they prevent oxidation. One, therefore, has to use a strong weapon to defeat them. Hydrogen peroxide is not strong enough for this; ozone, if enough is used, definitely is. This high reduction potential produced by ozone is commonly maintained for many months, even if no more ozone is added. This leads to a splendid growth of green algae.

Should the reduction potential be lowered by a weaker oxidation agent, red algae will begin to grow over night. If this equilibrium is maintained, the red algae slowly over-grow the green algae and the green algae die. In this case, which often happens in the aquarium, ozone is immediately added because the red algae are unable to break down the products of decomposition

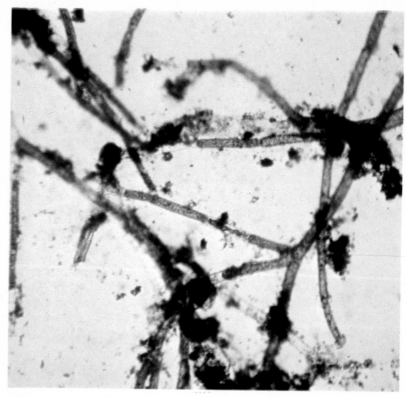

The normal respiration of fishes may be affected when the fine filaments of the green alga *Cladophora conglomerata* become enmeshed in the gills of fishes. Photo by Dr. H. Reichenbach-Klinke.

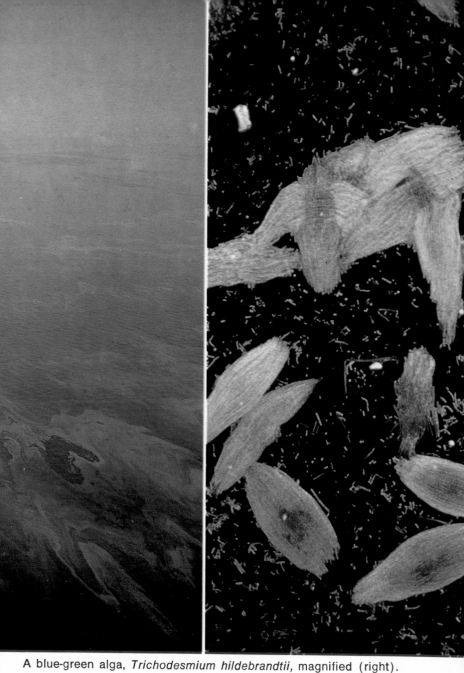

A blue-green alga, *Trichodesmium hildebrandtii*, magnified (right). Sudden blooms of this alga often occur in the areas about the Great Barrier Reef (left). Blue-green algal growth in the home aquarium is easily removed by reducing agents. Photos by K. Gillett.

of the green algae. This game can be repeated, with one and the same water, till one has grown tired of it. The actual oxygen content has no impact here: the continuous changes (red - green - red, etc.) take place at a small oxygen deficiency of 2 mg as well as at an excess of oxygen, provided it is molecular oxygen from the atmosphere. If we used atomic oxygen, we would forever move in the range of red algae! To begin with, the process is oxygen-dependent and a pure displacement of the electro-chemical equilibrium. But as soon as plenty of green algae are present there is no longer any possibility of oxygen deficiency. This is, however, not due to the well-known and striking phenomenon of additional oxygen being given off by the green algae, since the green algae grow because the reduction potential has reached a certain level, and this in turn would not have been reached had there been insufficient oxygen.

HIGH REDUCTION POTENTIAL

The growth of green algae always indicates a high reduction potential. The high reduction potential in turn shows that the water has reached the highest level of purity.

If one were to describe the process in a simplified way, by making the growth of green algae equal to clean water saturated with oxygen, a gross mistake might be made because any kind of disturbance can cause the reduction potential to drop very quickly, and at that moment there may already be a low or a high degree of oxygen deficiency. As has already been said, algae, used as biological indicators, respond slowly: green algae cannot disappear within a few hours.

But it is also a fact that sea water with green algae in it is surprisingly stable. To lower this high reduction potential, one would need either very strong reducing agents or very large amounts of weak reducing agents. Sea aquaria with an optimal growth of red algae are a different matter. Oxygen deficiency frequently occurs but need not necessarily be present. Relatively speaking, these waters are unstable. Minor changes in reduction potential already cause oxygen deficiency, and only very little more is needed to upset the balance completely. Sea-water with a growth of red algae, therefore, has to be handled with the greatest care; it is very sensitive.

BLUE-GREEN ALGAE

Blue-green algae apparently have a special aversion to low thresholds of oxidation. If the reduction potential is lowered, they promptly die. If it is lowered very carefully and slowly, they save themselves by moving up to the surface of the water. Undesirable blue-green algae can be destroyed with any reducing agent. For aquarium use, I can only recommend trypaflavine at 100 mg per 100 1. The application of other chemical reducing agents is either dangerous or extremely difficult, as they are oxidized too quickly in water. Methylene blue, itself a reducing system, is also suitable. Unfortunately, these convenient agents are dyes. The cell fluids of blue-green algae become blood-red through reduction. The process can be observed very well in sun light, especially at a high reduction with colorless quinol, resorcine, or others.

A coarse sand bottom like that shown under this pair of *Polycentrus schomburgki* contributes considerably to a better circulation in the aquarium when used with an undergravel filter and can be useful in the removal of hydrogen sulfide, a strong plant poison. Photo by R. Zukal.

Although these spawning *Capoeta oligolepis* are obviously in good health, the degenerated *Aponogeton* leaves in the background show that the requirements of the plants have not been completely satisfied. Photo by R. Zukal.

17.

Aquatic plants and water chemistry

In this chapter I would like to comment briefly on aquatic plants in general. I am not going to supply a list of species and their specific requirements—this can readily be found, and very beautifully illustrated at that, in BRUNNER: "Aquarienpflanzen". Only generalities connected with chemistry shall be mentioned.

It is a well-known fact that best growth and appearance are achieved above all when plants in an acid substrate receive an acid fertilizer; plants which prefer an alkaline reaction reach their optimum with alkaline to neutral fertilizers. In aquatic plants, a neutral to alkaline preference is absent. With the aid of comparative experiments, it is easy to prove that all popular aquatic plants have a striking preference for ammonium-nitrogen. In these experiments the control plants, which receive the same amounts of nitrogen and nitrate, all fall far behind. Aquatic plants obtain their nitrogen requirements with ammonium; the remainder—and that is most of it—is oxidized into nitrate. The nitrate store is not affected until the ammonium source has been exhausted.

In a tank stocked with fish, plant growth is always better if there are no additional sources of disturbance due to over-feeding. Apart from that, fish can influence plant growth favorably through expired carbon dioxide if there was a previous carbon dioxide deficiency because of too high a pH. If you, therefore, wish to virtually breed aquatic plants, you would consequently either have to put fish into a tank or to "force" the plants with ammonium (given in the form of ammonium sulphate). The addition of nitrate often recommended would only be a poor substitute solution. Into a new tank, we put sufficient ammonium sulphate to make up an ammonia content of 0.1 - 0.2

The presence of fishes in a planted tank influences plant growth favorably. The fishes in this tank are dwarfed by the lush growth of the plants, which obviously efficiently use the metabolic products of the fishes.

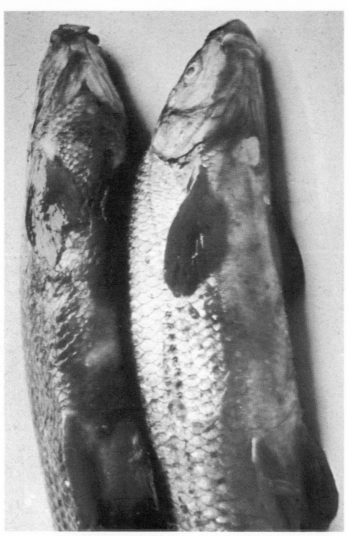

A tragic consequence of water pollution of unknown origin on fish.
Photo by Dr. H. Reichenbach-Klinke.

mg/1, and this at a pH of 6 - 7. Plants love the ammonium ion—not the ammonia! Naturally, this addition is not made when fish are present—not because the just noticeable ammonium content is harmful but because the fish themselves see to a constant dosage. Equally good plant growth can be achieved in a pure plant tank, in water which is completely devoid of nutrients, if a small bit of *Tubifex* worms is added daily or the tank is fertilized with urea at certain intervals. The agent which actually "forces" aquatic plants is, therefore, the ammonium ion.

Many natural waters supplied already have a quite considerable sodium chloride content so that many species can no longer be expected to show optimal growth. The chloride ion, important as a trace element, can become a strong source of disturbance if present in excess (200 mg or so). In that case, the plants are somewhat tinged with yellow and, above all, the leaves do not mature, although new leaves will continue to appear.

Ion antagonism has already been mentioned. To increase the salt content, plaster should preferably be used, not sodium chloride. High sulfate hardness is permissible, but not high carbonate hardness.

The strongest plant poisons are hydrocyanic acid (prussic acid) and hydrogen sulphide. While hydrocyanic acid never occurs in the aquarium, hydrogen sulphide, on the other hand, is produced all the time and is also "broken down" all the time —but not in the poorly aerated bottom layer. Aquarium water can be completely free of hydrogen sulphide—the aeration immediately gets rid of it— yet the bottom layer can, at any time, contain more of it than is good for a plant. A sure sign of hydrogen sulfide, which unfortunately is not noticed until the tank is being cleared, is the strongly smelling bottom layer which is of a grey-black color where zinc and iron are present or yellowish if metal ions are absent. This means that, under sulphur secretion, hydrogen sulphide was constantly being oxidized into water and sulphur. If these processes are not prevented altogether, they have to be confined to the filter—away from the roots of the plants! There is only one preventive measure, letting the bottom layer of the aquarium have its share of the circu-

Caulerpa is a common tropical marine alga, one of the few marine plants successfully kept in a well-managed tank. Photo by R. P. L. Straughan.

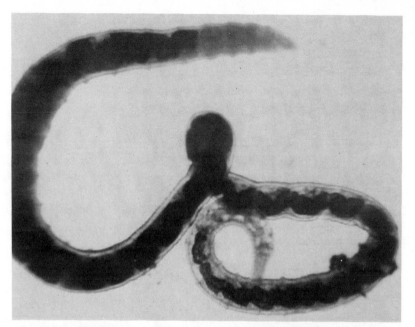

Small bits of *Tubifex* worms can be added as fertilizer in a tank containing plants only, but the addition of such organic materials can be prospectively very dangerous; commercially available plant stimulators are much safer. Photo by Dr. Rolf Geisler.

lation. Remember that a medium flow of water always has to go through from *below upwards,* never the other way around! After all, the bottom layer is not supposed to act as an additional filter!

Continued application of hydrogen peroxide leads to cell damage and chlorophyll destruction, but the maximum concentration would have to be maintained for at least 24 hours. There is no reason, therefore, why therapeutic treatment with hydrogen peroxide should not be carried out. As already mentioned, this amount is sufficient for the complete destruction of brown algae and diatoms. Whoever prefers to keep these algae should never use hydrogen peroxide.

ANALYTICAL SECTION

1.

Usage of chemicals and methods of analysis

Those who have no previous knowledge of chemistry are usually also afraid of chemicals. Such fear is totally uncalled for and should be replaced by knowledge and care. Never be careless, but don't become hysterical either. An example with regard to this is that the determination of carbonate hardness with 1/10 normal hydrochloric acid actually was "publicly condemned" with the advice that such dangerous substances should not be kept in the home. This 1/10 normal hydrochloric acid just mentioned consists of 0.36% hydrochloric acid!

Numerous chemicals, such as liquid ammonia, cleansers, and all types of stain removers, are stored in the home; these are by no means harmless and also have to be kept out of the reach of children. This goes for all medicines, too, and the so-called fish-therapeutic agents as well! So, if you are afraid of acquiring the chemicals mentioned, please also be consistent enough not to accidentally leave 10 pyramidon tablets openly lying around in a drawer—somebody, and this would be serious indeed, might swallow them by mistake. One definitely needs a drawer that can be locked. Nor does it make sense for the aquarist to be frightened of having hydrochloric acid in the house as long as the housewife herself has bottles standing about in some corner which are no less dangerous. The first rule, therefore, is: Take care!

1/10 normal solutions: They are used only for measuring purposes and have to be treated with care—they always have to be closed after use, and always with their correct top or cork. Any contamination later re-appears in the form of a gross measuring-error. If a pipette is dipped into a 1/10 normal solution, the pipette or burette has to be really clean; as rinsing-agents, these sensitive solutions are far too expensive. Never use a burette which has previously been dipped into another solution to take out 1/10 normal solution.

Concentrated solutions: Acid splashes on the skin are rinsed off with tap-water and a little soap. Alkali splashes have to be rinsed off with considerably greater amounts of water; it is best to neutralize with an acid such as vinegar, lemon, etc.

Nessler's reagent: Like almost all mercuric salts, it is poisonous. That it contains a strong alkali has to be remembered, too. Dishes and hands are cleaned under running water, and the basin is afterwards wiped. A broken clinical thermometer with a few scattered little balls of mercury in the room is incomparably more dangerous!

Potassium chromate: Poisonous heavy metal salt.

Universal indicators: Usually consisting of the very poisonous methyl alcohol as a solvent. It is precisely the indicator solutions the hobbyist likes to leave standing around near the aquarium!

Hydrogen peroxide: Concentrated 30% hydrogen peroxide is caustic. Dilute to 15% and store cool in a glass stopper bottle. Splashes on cloth bleach out the pattern.

Silver nitrate, 1/10 normal: Black spots gradually develop on the skin. If a garment has been splashed, it is gently wiped with tincture of iodine and rinsed with sodium thiosulphate; if necessary, repeat several times. The procedure does not damage the material.

Cer-IV-sulphate solution 1/10 normal: Cer is non-toxic, but the solution contains a considerable amount of sulphuric acid and is, therefore, caustic. Do not draw up with the mouth burette! Also causes bleaching spots on cloth.

GENERAL PRECAUTIONS

First of all, a plastic cover is put over the table or the piece of furniture used.

If a more highly concentrated hydrochloric acid is used indoors, the bottle is cooled thoroughly under running water before use; otherwise plenty of hydrochloric acid fumes would develop when the bottle is opened. They are harmless, but completely unnecessary, too, and, above all, steamed-up glass containers and shiny metal parts would become dull if frequently subjected to the fumes of hydrochloric acid. In all existing analytical procedures, the non-gaseous sulphuric acid could be substituted for hydrochloric acid—but sulphuric acid is really dangerous, its application is not advisable and *it does not belong in the home!*

METHODS OF ANALYSIS

Complications and time loss occur when accuracy or carelessness are applied in the wrong situation. In the analytical procedures below two words will re-occur at regular intervals; these are "about" and "exactly". Please adhere to them. If you read, for example: "The weighed amount is put into a measuring cylinder and filled up to exactly 100 ml", then "exactly" is not to be understood in the sense of a textbook of quantitative analysis. In this case one would, for instance, have to use a graduated cylinder or beaker and bring the liquid up to the temperature etched on to the retort. This precision would be nonsense for our purposes. "Exact", for us, only means that the burette or the measuring cylinder is held perpendicular, at eye-level, when read, in order to avoid distortion.

It is important that clean glass instruments and containers are used. Whether one uses an Erlenmeyer flask or a clean jam-jar instead, is quite immaterial. Whether, for the determination of the total salt content, one uses a manufactured burette or one's "own construction" does not have the slightest influence on the result. What actually matters in the individual tests is clearly emphasized in every case. Immediately after use, glass ware is rinsed first with plenty of tap water and then with distilled water.

If small quantities of substances are required, ten or a hundred times the amount is weighed and a concentrated solution of known content is prepared. This solution will then contain the desired amount, for example, per cc or per 100 cc.

For measuring purposes, one requires chemicals of "reagent" or "analytical" grade; otherwise "U.S.P." grade is sufficient. For regeneration purposes only, "technical hydrochloric acid" and "technical caustic soda solution" (sodium hydroxide, NaOH) are used.

In the analytical section, the term "milliliter", abbreviated to "ml", is used instead of "cubic centimeter" (cc). This abbreviation is etched on to the glass wares as well. That is: $1 \text{ cm}^3 = 1 \text{ ml} = 1 \text{ cc}$. The reagent bottles do not always bear uniform labels either—do not let this confuse you.

2.

Determinations of hardness and salts

TOTAL HARDNESS

Knowledge of the total hardness is important for the differentiation of sulphate and carbonate hardness. On its own, the figure is of no importance whatsoever, either with regard to fish or from a plant-physiological point of view. Most hardness tests are based on how difficult it is to obtain a permanent lather using a known amount of specially prepared soap solution. A known amount of aquarium water is placed in a stoppered bottle, and a standard (Clark's) soap solution is added. The amount of soap needed to make a lather which lasts 1 minute or longer is then compared to a table and the total hardness computed. Each kit used for hardness determinations will probably have its own directions, which should be followed explicitly.

CARBONATE HARDNESS

This measuring procedure involves the hardness substances which have an alkaline reaction.

Exactly 100 ml of sample are poured into a glass or porcelain container (or a cup which is white inside) and mixed with about 10 drops of methyl-orange indicator. The solution will turn yellow.

With a 5 ml burette we now draw up 1/10 normal hydrochloric acid and add this to the sample, drop by drop. Stir or shake the sample so that the acid is evenly distributed. As soon as the yellow color turns into orange-red, no more hydrochloric acid is added. 1 ml of 1/10 normal hydrochloric acid equals 2.8 degrees of German carbonate hardness.

Example: To make the color change from yellow to orange-red, 4.8 ml 1/10 normal hydrochloric acid were used. 4.8 by 2.8 = 13.44, or roughly 13.5 degrees (13.5°DH).

SULPHATE HARDNESS

We arrive at the sulphate hardness by deducting the carbonate hardness from the total hardness.

Example: The total hardness read was 18 degrees and the carbonate hardness 16 degrees. Sulphate hardness = 18—16 = 2 degrees.

Conversion values: 1 degree of German sulphate hardness equals a content of 24.3 mg calcium sulphate $CaSO_4$ per liter. If the hardness of a water is, however, to be raised by 1 degree sulphate hardness, 30 mg calcium sulphate $CaSO_4$ x 2 H_2O have to be added per liter. In crystal form calcium sulphate dissolves very well and does not form lumps.

TOTAL SALT CONTENT

Principle: To a strongly acid cation exchanger in H-form, an accurately measured sample of 100 ml is added. With the exception of carbonates, all salts are transformed into the appropriate free, non-volatile acid. The acid content of the sample leaving the exchanger is determined, in the presence of an indicator, with 1/10 normal caustic soda solution. Since the exchange takes place strictly quantitatively, the acid content is exactly the same as the total salt content when the carbonate hardness is added. Since the total salt content is only binding, physiologically, if bi- and univalent ions are considered separately, one cannot, logically, say the consumption equalled the total salt content, although, technically, this would be quite correct, more or less according to the pattern: "Total salt content equals . . . ml 1/10 normal caustic soda solution to 100 ml sample plus . . . degrees carbonate hardness."

We begin, therefore, by determining carbonate hardness and sulphate hardness (hereby including all bivalent ions). We then find out from the table below how many ml of 1/10 normal caustic soda solution are to be used up for the sulphate hardness we established and put down the excess—additional consumption in ml—as sodium chloride, also found in the table.

mg NaCl/liter	ml 1/10 normal caustic soda solution per 100	sulphate hardness in German degrees
14.6	0.25	0.7
29.2	0.5	1.4
43.8	0.75	2.1
58.5	1.00	2.8
73.1	1.25	3.5
87.7	1.50	4.2
102.3	1.75	4.9
117.0	2.00	5.6
131.6	2.25	6.3
146.2	2.50	7.0
160.8	2.75	7.7
175.5	3.00	8.4
190.1	3.25	9.1
204.7	3.50	9.8
219.3	3.75	10.5
234.0	4.00	11.2

Technical requirements: We use an exchanger burette with 50 ml resin, preferably the strongly acid cation exchanger Lewatit S 100 Gl with indicator. The resin rests on a nylon-wool swab and is covered with a little nylon-wool, too, so that it will not be whirled about. The glass tube is meant to prevent the column from running dry and attracting air. One could, of course, just as well draw up a plastic tube to the upper layer of the resin.

With a resin filling of 50 ml resin, the drop-speed may be about 2 drops per second at a layer depth of about 10 to 20 cm. The drop-speed is to be kept fairly constant so that the 100 ml of the added sample are to enter within 15 minutes. We then add 30-40 ml distilled water, again allow to drop up to resin level, and add a further 30-40 ml approx. Now the tap is turned on and we quickly rinse again with 100 ml distilled water. Through rinsing, the acid trapped in the resin by adsorption is meant to be washed into the sample. If we do not rinse with distilled water, the measuring result will generally be lower than it should be.*

Finally, all the liquid is mixed with methyl-orange indicator and, with a graduated burette or pipette, we add 1/10 normal caustic soda solution until the color changes from red to a light orange-yellow.

Example: 1. Total hardness 17 degrees, carbonate hardness 10 degrees, resulting in a sulphate hardness of 7 degrees. Looking up sulphate hardness of 7 degrees in the table, we find that at least 2.5 ml of 1/10 normal caustic soda solution are used up. If the actual amount consumed is no higher than that, the water contains hardness substances only. Result: Total salt content: Carbonate hardness 10 degrees, sulphate hardness 7 degrees, remaining salt content 0 degrees.

If the water contains other salts as well, there is increased consumption. From the table, this increased consumption is read as sodium chloride. An increased consumption of 1 ml 1/10 normal caustic soda solution then indicates a sodium chloride content of 58.5 mg NaCl/liter.

Example 2: To pure distilled water we add 15 mg sodium chloride per liter. In this case, of course, hardness equals nil, but after the exchange roughly 0.25 mg of 1/10 normal caustic soda solution would be required to neutralize the sample.

Sulphate hardness cannot, of course, be expressed as accurately, down to fractional parts of decimals, as listed in the table. One has to round off accordingly, usually upwards. The ml of caustic soda solution can in fact be read on the burette with up to two fractional parts of decimals.

TECHNICAL HINTS

The exchange burette is filled with distilled water and stored closed. If it is rarely used, the distilled water is topped up from time to time so that the resin continues to be immersed in water.

*We, therefore, use a glass tube with an internal diameter of 20-30 mm and a length of 25-30 cm. At 50 ml of resin, as suggested, half or two thirds of the column will be filled with resin.

With this preparation, depending on the salt content of the water, 100-200 tests can be carried out before regeneration. Regeneration is carried out when half the exchanger has been exhausted—this can be seen very clearly by the color change of the indicator resin.

For the regeneration of 50 ml resin we use 100 ml of diluted hydrochloric acid, analytical or reagent grade. About 20 ml hydrochloric acid (about 25%) are put into a graduated cylinder and diluted with distilled water to 100 ml. The acid is allowed to drop through, slowly. Catch it in a glass or plastic dish and neutralize with alkali or soda before throwing away! Rinse with 300 ml distilled water, using a very fast dropping speed. The rinsing water is added gradually and more is added when the liquid has dropped to resin level, etc.

CHLORIDES

Chloride content may be determined without also having to determine total salt content. Exactly 100 ml of the sample are put into a dish (white inside) or into a glass with a piece of white paper underneath. The sample has to be more or less neutral and is mixed with about 10 drops of a 10% potassium chromate solution as indicator. The burette (where the salt content is high, a 5-10 ml burette is required) is filled with 1/10 normal silver nitrate solution. While shaking the glass with the sample or stirring the sample in the bowl, gradually add the silver nitrate solution drop-by-drop until a red coloration can be observed. Previously, the sample had a bright-yellow color due to the potassium chromate indicator. Consumption in ml is read on the burette and the ml used are multiplied by 35.5. The sum arrived at is the chloride content in mg per liter.

Example: 1 ml was used for a sample of exactly 100 ml.
$$35.5 \times 1 = 35.5 \text{ mg chloride/liter.}$$

Conversion figures:
  1 mg/liter sodium chloride=0.607 mg/liter chloride.
  1 mg/liter chloride= 1.649 mg/liter sodium chloride.

3.

Determinations of nitrogen compounds

QUALITATIVE AMMONIA

Into a small white dish we put 5 drops of a 50% Rochelle salt solution, and then 10 ml of sample. We now add about 6 drops of Nessler's reagent and immediately afterwards about 6 drops of a 25% caustic soda solution. If after a short period (1-2 minutes or more) the liquid is slightly tinged with yellow, the water contains roughly 0.15-0.25 mg ammonia per liter (at a low pH, an equal amount of ammonium). If there is instant yellowish coloration, about 0.5-1 mg are present per liter. At a brown coloration, the ammonia content lies above 3 mg/liter.

Sea-water: Put at least 10 drops, preferably 15 drops, of 50% Rochelle salt solution into a test-tube, add the sea-water sample and immediately afterwards about 6 drops of Nessler's solution. The addition of Nessler's will cause a strong milky turbidity; without shaking the sample, add about 6 drops of 25% caustic soda solution, then shake very gently. Now the solution will become quite clear again, but after about 10 minutes it will again become cloudy. That everything is added in the correct sequence is strictly necessary when sea-water is being tested; it is also of importance that all additions are made promptly, one after the other. You must work without interruption, since you only have 10 minutes to run the test. Colors are read as in the freshwater test.

QUANTITATIVE AMMONIA

For this, we require a control solution (see p.112 for preparation) and two small white porcelain dishes of the same shape, not just of the same volume, as the colors must be matched at more-or-less equal depths.

Into the first small dish we put Rochelle salt solution, then the sample plus Nessler's reagent and alkali, as in the qualitative ammonia determination. Into the second dish we also put Rochelle salt solution, followed by 10 ml distilled or tap water and the same amount of Nessler's and alkali. A 5 ml burette is filled with control solution which is added, while shaking or stirring, drop by drop in quick succession until the colors become identical. The ml of control solution needed equal mg ammonia per liter.

Example: With its markedly yellow coloration, the sample was positive. To the control sample 2.5 ml control solution had to be added so that the colors became identical: Ammonia = 2.5 mg/liter.

Please note: Never measure beyond 3 mg/liter. If there is a higher content, the sample has to be diluted and the dilution factor to be taken into consideration.

Dilutions:—

To 5 ml sample add 5 ml distilled water and measure again. The result has to be multiplied by 2.

To 0.5 ml (to be measured accurately, with a measuring pipette!) add 9.5 ml distilled water. Multiply the result by 20.

To 1 ml add 9 ml distilled water. Multiply by 10.

QUANTITATIVE NITRATE

Principle: The nitrates are reduced to ammonia with zinc powder and acid and the ammonia content is determined quantitatively. Through multiplying by the factor 3.64, we obtain the nitrate content in mg/liter. Or, to put it another way, by the process of reduction, 1 mg ammonia is formed from 3.64 mg nitrate.

To about 15 ml sample, we add a small knife-point of analytical grade zinc powder and about 4-5 drops of concentrated hydrochloric acid and let it stand for 15-20 minutes. We require four small dishes, of uniform shape, with a volume of about 12-15 ml.

Into the first dish we add to exactly 5 ml distilled water, 5 ml of the reduced sample; into the second dish, 9 ml distilled water and 1 ml sample; into the third dish, 9.5 ml distilled water and 0.5 ml of the reduced sample. Rochelle salt, Nessler's, and alkali are added as in the ammonia determination, as if each dish was the original sample. Choose the dish with the weakest yellow color and adjust its color with control solution to that of the fourth dish, which contains distilled water and chemicals.

Example 1: Dish 1, sample of deep-brown coloration, dish 2, sample light-brown, sample in dish 3 pale-yellow. To make the colors identical, 3 ml control solution were required; this equals ammonia content of 3 mg/liter by dilution factor $20 = 60$ mg ammonia by $3.64 = 218.4$ mg nitrate per liter, roughly 220 mg.

When testing nitrates in sea-water, some slight deviations from the basic test are necessary. Before diluting, add more hydrochloric acid to redissolve any zinc precipitations. Rochelle salt, Nessler's, and sodium hydroxide are placed in an empty dish, and *then* the diluted sample is added.

The nitrate content of sea-water is higher as a rule—the dilution of 0.5 to 9.5 is, therefore, not always sufficient. We then prepare another dilution of 0.5 ml sample plus 9.5 ml distilled water, take half of this (5 ml), and add another 5 ml of distilled water. Equally well one could, of course, add 19.5 ml distilled water to 0.5 ml of the reduced sample, pour out 10 ml and use the remaining half for measuring. Choose whichever method you prefer. In both cases, the dilution factor is by 20 by 2= by 40.

Example 2: None of the three diluted samples has a yellow color. This means that the dilution consisting of 5 ml reduced sample plus 5 ml distilled water was already much too strong, the nitrate content must lie below 1.8 mg/1. If the result is still of interest, one can determine the undiluted sample, but less than 0.9—1 mg nitrate/liter can no longer be measured. With this method, the measuring range thus begins with about 1 mg/1 nitrate and by diluting can be extended upwards as desired; but one must not forget to include the dilution factor in the calculation. If one wants to measure from 1 mg nitrate per liter upwards, he cannot get around the preparation of dilutions whichever other method (and there are many), of nitrate determination is used.

Ignore the decimal point and round-off to whole numbers. With good illumination, the inaccuracy of this method will be ± 10%, at twilight rising to up to 50%.

SOLUTIONS FOR NITROGEN DETERMINATIONS

Ammonium chloride 3% solution is prepared by weighing three g of analytical grade ammonium chloride and diluting it to 100 ml. This is the standard or stock solution which will keep for about a year in a well sealed bottle.

Control solution for the quantitative determination of ammonia. Take 0.5 ml of the 3% ammonium chloride solution and add it to 500 ml distilled water. Keep about 100 ml of this solution; pour the rest out. A new supply is prepared every other month or so. 1 ml of this control solution to 10 ml sample indicates 1 mg ammonia per liter.

Since the standard and control solution is a measuring substance, really clean and accurate work is required. Use new medicine amber bottles with new stoppers—rinse in distilled water before use. Measuring cylinder and burette also have to be really clean—above all, no traces of alkalis must be around!

Preparation of the 50% Rochelle salt-solution: 50 g (approximately) of analytical Rochelle salt are dissolved in 100 ml of distilled water. After stirring for some time, the salt slowly begins to dissolve. We now add 3 ml (or 60 drops) of Nessler's reagent, stir again quickly, and let stand over night. The following day the solution is poured into a bottle—any sediment present is left behind. Slight brown coloration does no harm. This solution is stored in a brown dropping bottle as should any reagent which is used only in drop form.

With regard to Nessler's reagent, the following should be observed: make sure you order Nessler's reagent solution A and Nessler's reagent solution B (equal amounts). A mixture of two equal parts gives us the reagent, ready for use. Stored separately, the solutions keep indefinitely. A prepared, complete mixture, if still offered anywhere, should not be purchased.

4.

Miscellaneous determinations

ATOMIC (NASCENT) OXYGEN

A rapid method is only possible with 1/10 normal potassium permanganate solution or with 1/10 normal Cer-IV-sulphate solution. Determination with potassium permanganate is old-fashioned and very complicated since the solution does not keep; it would have to be accurately "adjusted" before each use —its content would have to be checked. Cer-IV-sulphate solution, 1/10 normal, may be considerably more expensive but can be kept for years without changing. The high price saves a lot of trouble and annoyance!

Put 100 ml of sample into a pint (500 ml) container and add 10 ml analytical grade 25% hydrochloric acid and about 10 drops of the reduction indicator, Ferroin. The sample becomes bright-red in color. To draw up the 10 ml of hydrochloric acid, do not use anything but a 10 ml pipette with bulb or other suction device. A graduated 1 ml burette is then filled with Cer-IV-sulphate solution, 1/10 normal, and, while gently shaking the glass, the solution is added drop by drop until the color changes from red to sky-blue.

At the dosage suggested under "Oxygen", after adding 1 ml of 15% hydrogen peroxide to 20 l aquarium water (about 1 drop per liter), 0.55 ml Cer-IV-sulphate 1/10 normal to 100 ml test-sample would be used.

The first 0.15 ml Cer-IV-sulphate solution are regarded as "ticking-over", that is they are used up in any case.

In practice, you can do without the burette. The Cer-IV-sulphate solution is then poured into a dropper bottle and the drops used are counted. In that case, however, the number of drops depends on the size of the dropper, too. If the dropper has a large opening, 6-8 drops will be required at the dosage recommended, and of these the first 2-3 drops will definitely be needed to reach zero-position. If you wish to work without a burette according to the drop-counting method, the drop-requirements are tested by preliminary measuring as follows:

1. You measure how many drops are required for the color change with this dropper when no peroxide has been added to the water; this number of drops is kept in mind for all future reference.

2. The amount of hydrogen peroxide mentioned, accurately calculated, is added to the aquarium and after a short period (allowing it to be distributed in the tank), measure again. One then knows that at the correct dosage drops are required to 100 ml sample (6-7 or 8 drops).

If measuring is repeated a few hours later, we use fewer drops, depending on how severely the water had been contaminated. If finally, when measuring yet again later on, only as many drops are used as had been estimated for the zero-position, no further additional oxygen is present. Measuring at intervals of 2 hours or so, of course, only makes sense if one wants to determine the degree of water purity from the decreasing of the oxygen content (additional atomic oxygen) within a certain length of time.

Last but not least, this measurement is important in emergencies, such as when the first addition of hydrogen peroxide has been effective instantly but its intensity has already markedly dwindled after a short period (half an hour to an hour and a half). If you want to add another lot of hydrogen peroxide a little later, check, first of all, whether the first quantity has in fact been decomposed. Although a once-only overdose of 100% is harmless, it is dangerous to keep repeating the dosage without checking.

INDICATIONS OF WATER PURITY

The greater the organic or reductive content of a water, the faster the content of atomic oxygen disappears. In aquaria which are dangerously polluted, the content vanishes after 2 hours or, in extreme cases, after 30 minutes. If after 2 hours the consumption has gone down by half (for instance 5-6 drops of Cer-IV-sulphate instead of 8 drops, since the first 2-3 "ticking-over" drops do not count), the water is already quite heavily contaminated. With very clean water, a slight excess (1 drop above "ticking-over") still has to be present after 6-7 hours.

The more brown algae and diatoms water contains, the faster the excess peroxide disappears. Green algae do not have any notable influence on the speed of decomposition and do not, therefore, upset the measuring—unless they are dying.

Instead of methyl-orange indicator, Töpfer's reagent can also be used. When measuring the carbonate hardness, the measuring-inaccuracy otherwise common is practically nil since this indicator does not react to carbonic acid. The color change, too, is more marked.

SEA-WATER pH

1 g a-naphtholphthalene is dissolved in 400 ml of 96% alcohol; to this we add 100 ml of a 2% alcohol phenolphthalein solution. Presumably one could just as well use denatured alcohol instead of the expensive ethyl alcohol. The result is the same, but the keepability has not been tested.

pH-value below 7.5	almost colorless	beige (tinged with yellow)
pH 7.5-7.8	green	(green)
pH 7.9-8.1	pure turquoise	(sea-green)
pH 8.2-8.3	sky blue	(pale-blue)
pH 8.4-8.5	dark blue	(dark-blue)
pH 8.6-8.7	blue violet	(pale violet)
pH above 8.7-9	pure dark violet	

With this mixture, not only the color changes in the way described above, but reflection ability and color intensity also change. It is quite impossible (even by artificial illumination) to confuse turquoise with green and sky blue, since the color changes do not take place gradually; turquoise appears less dull than green.

If a teaspoonful of sodium bicarbonate (U.S.P.) is dissolved in a glass of distilled water at room temperature and aerated with the air pump for 30 minutes, we have a standard solution of pH 8.7-8.75. An accurate standard solution of 8.4 is produced by dissolving 8.4 g of sodium bicarbonate in 1 liter of distilled water and then aerating the sample.

QUININE

Quinine is used in the treatment of certain fish diseases. It is poisonous and, therefore, should not be allowed to remain in the water. The exchanger resin Lewatit MP 60 is suitable for quinine removal if its range is slightly expanded through being made mildly acid with 0.5-1% hydrochloric acid. It is important to check afterwards whether all the quinine has, in fact, been filtered out of the water.

To test for the presence of quinine, add a small pinch of wolframatophosphoric acid (tungstophosphoric acid) to a 20-30 ml sample. When present, quinine causes mild turbidity; at a content of 1 g of quinine hydrochloride in 100 1 water, we observe a yellowish-milky opalescence.

APPENDIX

I.

Ion exchange device

An apparatus of very reasonable cost can be produced from two 5-liter plastic buckets fastened together, with a screwed-in bottom sieve. The bottom sieves have to be specially made by a synthetics manufacturing company. Into the bottom of the bucket a ring is put first of all, then on top of this the perforated plate with the nylon sieve which is secured with another ring. Between the bottom of the bucket and the sieve cover, the lower ring makes a small but sufficient hollow space which ensures that the water flowing through can pass the resin evenly without forming canals or being diverted.

We require one 5-liter bucket which is slim and tall and another which is less conical. This second bucket is put into the first and the overlap of about one third is cut off with a sharp knife. Both buckets are equipped with a bottom sieve, screwed on to the bottom of the buckets with plastic screws. The trimmed bucket is provided with numerous holes so that the draining water can run off through a very broad area. The lower bucket gets one single drainage hole with a diameter of about 5 mm.

Into the upper bucket we put the strongly acid cation exchanger in H-form, into the lower bucket the mild to medium-basic anion exchanger in OH-form. For regeneration, the containers are simply pulled apart.

One could, of course, manage without a bottom sieve by boring holes into the bottom of the bucket and storing the resin on a layer of nylon wool. Although this does not cost anything, the disadvantages should be pointed out, too: if, for instance, solution is poured in during regeneration and rinsing, the resin which is light anyway will be whirled up and then the nylon wool will come up, too—consequently, the resin will, of course, escape through the holes! At the small resin quantity of 2-3 liters, the depth of the layer should really be three to four times larger than the diameter. This is, of course, not so when we use the bucket method. But the error caused by this is of no impact here. An exchanger combination of the same capacity made of plastic or some other synthetic material would cost much more, even without resins.

Into the lower bucket we put about 3 liters of mild or medium-basic anion exchanger, into the upper bucket 2-2.5 liters strongly acid cation exchanger. At a medium salt content, 400-500 liters distilled water can be produced with this combination; then we have to regenerate.

Where partial salt removal is all that is required, one bucket with a strongly acid cation exchanger and a filling of about 4 liters resin are sufficient. With this quantity, 600-1000 liter water with a medium salt content (equalling a hardness of 15-25 degrees) can be rendered partially salt-free. Aquarists having a water with a low total salt content but too high a carbonate hardness would be at a particular advantage here. For instance: Sulphate hardness 1-2 degrees, carbonate hardness 4-6 degrees, other salts 10-30 mg/liter. From one regeneration to another, they could then carry out partial salt removal for 2-3 cubic meters water or total salt removal for 1-2 cubic meters.

On top of the resin in the upper bucket we put a plastic bowl of suitable size with many small holes to prevent the resin from being whirled up when the combination is held under the tap. This also protects the surface from access of air, and the resin will not dry up. If the exchanger is used only at long intervals, it is wrapped up in a plastic cover. This is of particular importance if one goes away for a long period. But the combination can also be put into a larger bucket and kept immersed in water; the main thing is that the resin does not become dry.

The combination is exhausted when the drainage has a pH of below 5 or above 6.5. If the pH is below 5, for instance 3.5-4, the cation exchanger can still be used but the anion exchanger is exhausted. If the pH is roughly that of the tap-water running in, the cation exchanger is definitely exhausted and the mild to medium-basic anion exchanger no longer functions either. After having used the combination for some time, one should, therefore, carry out a quick pH check of the drainage water by dipping in a piece of litmus paper.

II.

Regeneration of ion exchangers

REGENERATING—THE CATION EXCHANGER

For 2.5 liters resin, use 5 liters of 10% hydrochloric acid. Hang up the exchanger bucket and put an empty plastic bucket (10 liters) underneath. Apply regenerating acid and catch it in the lower bucket. But first of all the exchanger bucket is filled with tap-water to prevent a spontaneous contact of the resin with hydrochloric acid. When the 5 liters of acid have been applied, the bucket is left hanging for 10 minutes and then rinsed with three to four liters of tap-water. Into the now acid-filled bucket we put about one liter of technical caustic soda solution, and the neutralized drainage is thrown away. The emptied bucket is put underneath again to catch another 10 liters of rinsing water which also has to be neutralized with very little caustic soda solution before it is poured out. We then carefully rinse everything we have used before we put it away.

5 liters of 10% hydrochloric acid are prepared by pouring 4 liters of tap water into a bucket and adding a liter of concentrated technical hydrochloric acid. Hold the bottle in the bucket and cover with a sheet of newspaper while pouring. Work with rolled-up shirt-sleeves and rinse with running water any acid splashes on the skin. It is important to protect the face, clothes, and chromium-plated faucets, etc. The hands are not harmed by the acid if they are rinsed afterwards!

REGENERATING THE ANION EXCHANGER

These resins have a lower filter resistance, especially Lewatit MP 60. The anion-exchanger bucket has to be put in a plastic bowl so that the diluted caustic soda solution poured in can slowly flow through. The bucket with the plastic bowl can, however, also be put into a sink or the bath so that the drainage lye need not be collected in the bucket. Afterwards we then wash the lye out of the bath with plenty of water. Regeneration is carried out with 3% caustic soda solution prepared with acid water from the cation exchanger. For 3 liters of resin we use 6 liters diluted lye, prepared by mixing 300-350 ml technical caustic soda solution with about 5.5-6 liters water. When pouring, cover with newspaper. Rinse with acid water from the cation exchanger. 6 buckets of acid water will be required (sixty liters), possibly more. The acid water is added gradually —wait until three to four liters rinsing water have gone completely through, then add more while making sure that the resin is stirred up by the in flowing water. Rinse rapidly—60 liters rinsing water can be poured through within 15-20 minutes. The rinsing process is completed when the drainage water no longer has an alkaline reaction.*

It takes about one hour to regenerate the whole combination, or possibly 2 hours the first time before one gets the hang of the procedure. If you work on a large scale, you should, therefore, decide beforehand whether it would not be wiser to get a combination with two 10-liter buckets at the beginning. If you, for instance, possess several larger aquaria and have a water supply with an extremely high salt content (sulphate hardness 30 degrees, carbonate hardness 10-15 degrees, sodium chloride 200 mg/liter, other salts such as sodium sulphate, sodium bicarbonate 100-150 mg/liter—such water really exists!), you should seriously consider this. 10-liter buckets are not all that expensive and the regeneration of the larger set does not take much longer than that of the smaller one. Apart from the larger amounts of resins, only the consumption of chemicals will be higher, and that is of no consequence anyway.

---

* The exact amount of rinsing water required depends on the type of resin. If the concentration of the regenerating agent and the rinsing method remain constant, the consumption of rinsing water for the type used also remains constant. If the rinsing water is added step by step (about 2 liters each time, then allowing to drain, another 2 liters, etc.), total consumption can be further reduced.

REGENERATION WITH CITRIC ACID

The exhausted cation exchanger is pre-regenerated with a concentrated sodium chloride solution which is poured over it. Afterwards it is quickly rinsed and then regenerated with 20% citric acid instead of hydrochloric acid. The acid drainage water can be poured out without neutralizing with lye. The resins are then rinsed with tap water. For larger exchangers this method is, of course, rather expensive. Pre-regeneration with sodium chloride is essential for the citric acid method. If it is left out, the resin mass can become brittle because of the barely soluble earth alkali citrates and then still has to be treated with diluted hydrochloric acid.

If regeneration with hydrochloric acid is replaced by regeneration with citric acid, a liter of water may be at least 50% cheaper than purchased distilled water, and the caustic soda solution could safely be replaced with a 6% soda solution, but distilled water produced by normal regeneration costs next to nothing and can be used quite liberally—cubic meters of it.

If, for any reason, you should wish to use strongly basic anion exchangers, the following must be noted.

Strongly basic anion exchangers have to be run in carefully; that is, they have to operate at a minimal level for a long period. They should also be a few months old before use. They contain trimethylamin, added on production, which is difficult to wash out; trimethylamin is fish-poisonous.

Strongly acid and mild or medium-basic resins also have to be "run in", but with them it is much simpler. Factory-new resins are regenerated and the exchanger hung under slightly running tap over night. This is repeated once more after regeneration and the exchanger is then ready to be used for aquaristic purposes.

III.

Ion exchange resins

The following list only contains resins which have been thoroughly tested for aquaristic suitability. Resins which are not suitable for aquaristic purposes have been labelled accordingly.

Strongly acid cation exchanger: Permutit RS; Lewatit S 100, G 1 with colored indicator; Amberlite IR 120. Temperature tolerance: up to 110 degrees; pH range constant: 0-14.

Mildly acid cation exchanger: Lewatit CNO. Temperature 40 degrees, pH range 0-9; Amberlite IRC-50, pH range 0-14 = best mildly acid resin; Permutit C = aquaristically unsuitable.

Mild and medium-basic anion exchanger: Lewatit MP 60, temperature up to 100 degrees, pH 1-14; Amberlite IR 4 B = aquaristically unsuitable; Permutit E 7 P, maximum temperature 40 degrees, not for sea-water.

Strongly basic anion exchanger: Lewatit M 500 G 2, with indicator, temperature up to 70 degrees, pH 1-14; Permutit ES, temperature up to 40 degrees (easy to run in); Permutit ESB, particularly strongly basic, aquaristically of no use whatsoever; Amberlite IRA 410, temperature up to 40 degrees, pH 1-14. Another excellent strongly acid cation exchanger with indicator is Serdolith red.

INDEX